Current
Topics in
Biochemistry
1973

ACADEMIC PRESS RAPID MANUSCRIPT REPRODUCTION

Based on a series of lectures held at the
National Institutes of Health
Bethesda, Maryland
during March 1973

Current Topics in Biochemistry 1973

Edited by

C. B. ANFINSEN
ALAN N. SCHECHTER

National Institutes of Health
Bethesda, Maryland

ACADEMIC PRESS, INC. 1974 New York and London

ACADEMIC PRESS, INC.
111 Fifth Avenue, New York, New York 10003

United Kingdom Edition published by
ACADEMIC PRESS, INC. (LONDON) LTD.
24/28 Oval Road, London NW1

LIBRARY OF CONGRESS CATALOG CARD NUMBER: 72–77222

PRINTED IN THE UNITED STATES OF AMERICA

CONTENTS

CONTRIBUTORS

Henryk Eisenberg
Fogarty International Center, National Institutes of Health, Bethesda, Maryland and Department of Polymer Research, The Weizmann Institute of Science, Rehovot, Israel

Donald S. Fredrickson
Molecular Disease Branch, National Heart and Lung Institute, National Institutes of Health, Bethesda, Maryland 20014

David H. Sachs
Immunology Branch, National Cancer Institute, National Institutes of Health, Bethesda, Maryland 20014

Harold A. Scheraga
Department of Chemistry, Cornell University, Ithaca, New York 14850

Robert T. Simpson
Section on Developmental Biochemistry, National Institute of Arthritis, Metabolism, and Digestive Diseases, National Institutes of Health, Bethesda, Maryland 20014

E. Brad Thompson
Laboratory of Biochemistry, National Cancer Institute, National Institutes of Health, Bethesda, Maryland 20014

Bernhard Witkop
Laboratory of Chemistry, National Institute of Arthritis, Metabolism, and Digestive Diseases, National Institutes of Health, Bethesda, Maryland 20014

PREFACE

This volume is based on a series of lectures held at the National Institutes of Health. This group of lectures is the most recent in a program, which was originated in the mid-1960s, to review various research fields for the scientific community at the Institutes. The topics for these series were chosen to emphasize and summarize active fields of general interest for a diverse audience of scientists. The speakers were encouraged to present an overview of their fields rather than a detailed discussion of current research problems.

The substantial attendance at these lectures, and the response to the two series which have been published in this format, have reinforced our belief that such 'state-of-the-art" reviews are useful to a large number of research workers.

As in the previous published series, this collection covers a wide range of topics, including studies of pure proteins, gene expression in eukaryotes, and metabolic diseases. The unifying theme is the use of chemical methods in studying biological problems.

We again thank Mrs. Anne Ettinger and Mrs. Dorothy Stewart of the Laboratory of Chemical Biology, NIAMDD, for their skilled help in the assembly of the book and we thank the staff of Academic Press for its cooperation.

C. B. Anfinsen
A. N. Schechter

Current Topics in Biochemistry

1973

PREDICTION OF PROTEIN CONFORMATION

Harold A. Scheraga

Department of Chemistry
Cornell University
Ithaca, New York 14850

I. Historical Introduction

In this lecture, I will present a summary of the
present status of the problem of predicting the conformation of a protein from a knowledge of its amino
acid sequence, and will also mention briefly some preliminary results on the calculation of the preferred
conformations of enzyme-substrate complexes.

It is of interest to begin with some historical
perspective of the problem. Almost 25 years ago, when
Sanger and his collaborators deduced the first amino
acid sequence of a protein, insulin (58),--at a time
which preceded the determination of the first crystal
structure of a protein, myoglobin (32) and hemoglobin
(47),--we began to try to determine the structure of
a protein in solution by chemical and physico-chemical
methods. Our approach was to find the location of
many local pair-interactions which would act as constraints on the folding of the polypeptide chain. The
covalent structure already provided a knowlege of a
few interactions, *viz.*, the specific half-cystine
residues involved in disulfide bonds, and we attempted
to deduce the location of non-covalent interactions
between specific residues. Having available the amino
acid sequence of insulin (58), we began with this protein (60, p. 241). However, for a variety of reasons,
primarily its insolubility in the neutral pH region,
insulin proved to be a difficult protein with which to
work. Therefore, with the knowledge that Stein and

1

Moore and their collaborators (29,66–68) at the Rocke-
feller Institute and Anfinsen (54) and others (21) at
the National Institutes of Health were working on the
amino acid sequence of bovine pancreatic ribonuclease,
a protein (and, incidentally, an enzyme) with physical
properties much more compatible with our experimental
approach, our efforts were turned toward ribonuclease
(60, p. 270), whose covalent structure is shown in
Fig. 1, and also, for similar reasons, to hen egg white
lysozyme (60, p. 254).

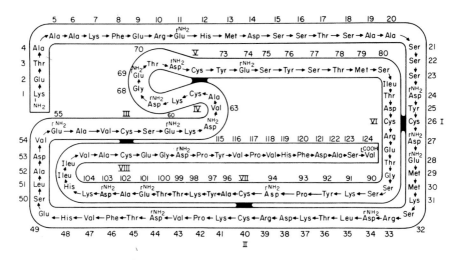

Fig. 1. Amino acid sequence of bovine pancreatic ribonuclease A (21,29,54,66–68).

Over a period of about 10 years, during which most
of our efforts were devoted to ribonuclease, we acquired information about the proximities of several
residues (61), in addition to the known location (68)
of the four bridges between cysteine residues. In particular, three of the six tyrosyl residues were paired
specifically with three of the eleven carboxyl groups
(all aspartyl residues). Considering that there are
3300 ways to pair 3 of 6 tyrosyls with 3 of 11 carboxyls
(61), this specific pairing [which is consistent with

the subsequently-determined X-ray structure of ribonuclease (31,78)] represented the fruitful results of a
long series of chemical and physico-chemical studies
(61). In the same period, it had been demonstrated
that His-12 and His-119 were near each other (5,22,26)
and close to Lys-41 (25,28), all three constituting
part of the active site. Having available these constraints (4 disulfide bonds and 3 tyrosyl-aspartyl
interactions, and the proximity of His-12, His-119,
Lys-41), we began to consider how these might be used
to determine the three-dimensional structure of the
whole protein. At this point, an assumption was introduced, *viz.*, that the native conformation would be the
thermodynamically most stable one. Thus, one would
make use of the above constraints and try to find the
conformation of lowest free energy. The validity of
this assumption gained considerable support from the
experiment of Anfinsen (4), who showed that (upon re-
oxidation) reduced ribonuclease (with its 4 cystines
converted to 8 cysteine residues) could fold spontaneously to yield the native structure with formation
of the correct disulfide bonds. With this background,
we began (44) to develop the methods to generate an
arbitrary conformation of a protein and compute its
energy, and ultimately its free energy, subject to any
constraints such as those discussed above, so that one
could select out of the enormous number of conformations
accessible to the polypeptide chain the one which corresponds to the lowest free energy.

II.   Nature of the Problem

To state the problem in another way, one can imagine having a ball and stick (non-space-filling) model
of a protein like ribonuclease, *i.e.*, of a connected
sequence of amino acid residues (with the disulfide
bonds in their proper places). Since the lengths of
the sticks are fixed and the holes are drilled in the
balls in specific places, the bond lengths and bond

angles are fixed[1] at values set by the manufacturer. Therefore, the only degrees of freedom (in order to change the conformation) are the dihedral angles for rotation about single bonds (with the amide groups maintained in the planar trans conformation[1]). Remembering that a protein of the size of ribonuclease has about 500 single bonds in its backbone and side chains, it is easily seen that, by such rotations, it is possible to generate millions and millions of conformations, only one narrow range of which corresponds to the native protein. In computational language, the dihedral angles are the independent variables for generating any arbitrary conformation, and the energy of each conformation is computed in a search for the conformation of lowest energy (56,59,62); various entropy contributions are included to obtain the free energy (62). It is now possible to minimize the total interaction energy, including *all* pairwise interatomic interactions, with respect to the dihedral angles for a protein of the size of ribonuclease in a reasonable amount of computer time (75). While such computations have been, and are being, carried out, I want to concentrate in this lecture on the as-yet-unsolved problems which must be surmounted before we can predict the three-dimensional structure of a protein solely from a knowledge of its covalent structure.

III. Empirical Energy Functions

Without getting involved in mathematical details, I simply want to say that we have available empirical

---

[1]In a computation, one can vary the bond lengths, bond angles, and planarity of the amide group by introducing appropriate force constants (45,77). While we have allowed for such degrees of freedom, most of our computations have been carried out with fixed bond lengths and bond angles and planar trans amide groups (*i.e.*, rigid geometry), selected separately for *each* type of amino acid.

energy functions, based on pair interatomic interactions, for carrying out such computations, and also procedures for including the effect of hydration. In the absence of solvent, the parameters of the empirical energy functions have been refined (42) by computing the lattice constants, intermolecular binding energies and some intermolecular force constants of a large number of crystals of small molecules, such as, hydrocarbons, carboxyl acids, amides, etc. These functions have also been tested on a variety of model systems. For example, we have computed the preferred conformations of the N-acetyl N'-methyl amides of the twenty naturally occurring amino acids (38), including not only the energy but also the librational entropy in the calculations. The computed relative amounts of the two dominant species, a five-membered and an equatorial seven-membered hydrogen-bonded ring, respectively, agree in general with values deduced from infrared and nuclear magnetic resonance measurements on these compounds in nonpolar solvents[2] (38). As another example, the correct helix sense of a large number of α-helical homopolymers has been computed with these parameters (43). Thus, while there is always room for improvement, it appears that we have available a reasonably reliable set of energy functions for carrying out computations on polypeptides and proteins in the absence of water. While a procedure is available (16) for including the role of the solvent, and has been applied in a number of computations [e.g., the formation of a hairpin turn in a long α-helical section of poly-(L-alanine) in water (65)], we are devoting considerable

---

[2]Recently, by considering the theoretical results (38) and related experimental data, it has been shown (10) that a third conformation (with no internal hydrogen bond, and designated as a γ conformation; see Fig. 2) can exist to a large extent for amino acids other than glycine. Heretofore (38), the γ and equatorial seven-membered ring conformations have been considered to be the same.

attention at present toward the improvement of the treatment of the solvation of a polypeptide chain.

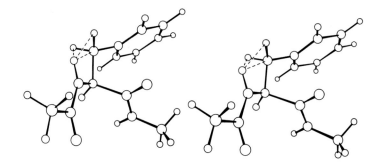

Fig. 2. ORTEP stereo diagram of the dipeptide N-acetyl N'-methyl-L-phenylalanine amide in the γ conformation (ϕ =-60°, Ψ = 140°, χ_1 = 180°, χ_2 = 80°) (10). The dashed lines indicate non-bonded interactions which are thought to affect the N-H stretching frequency.

IV. Multiple-Minima Problem

To turn to the as-yet-unsolved problems, let us recall that a polypeptide of 100 residues has about 500 independent degrees of freedom[1] counting dihedral angles of rotation about single bonds: two per residue in the backbone and an average of about three per residue in the side chain. Thus, the energy surface is a 500-dimensional one and is very complex. To illustrate the difficulty that arises from this complexity, let us pretend, for the sake of drawing a sketch in two dimensions, that the energy, E, is a function of only one variable, q. This dependence might appear like the curve shown in Fig. 3. From this diagram it becomes obvious that conventional minimization procedures will lead to the minimum in the *same* potential energy well as the conformation from which the computation was started. What is required is a procedure to surmount intervening potential barriers (of course, in the 500-dimensional space) in order to reach the *global* minimum, the one of lowest energy or, to reach

the minimum corresponding to the native protein, if it
is not the global minimum. The rest of my lecture will
be devoted primarily to a description of our efforts
to overcome this as-yet-unresolved difficulty, and of
the information we have gained during the course of
this work about the factors which determine the preferred conformation of a polypeptide chain.

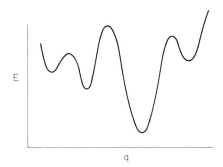

Fig. 3. Schematic two-dimensional representation of
the energy as a function of conformation.

Our initial efforts were devoted to the development of mathematical procedures (11-13,17,18) to at-
tain the global minimum. These procedures have proven
to be successful for small oligopeptides, but would
consume too much computer time for a larger structure.
The largest structure to which such procedures have
been applied is deca(L-alanine) in water (17). Starting with an α-helical conformation (a local minimum),
this decapeptide was carried through 14 successive
lower minima (with a net decrease in energy of 38 kcal/
mole, without yet reaching the global minimum), the
conformation at each successive minimum departing more
and more from the initial α-helical structure. It is
our view, at the present time, that suitable algorithms
are not yet available to provide a mathematical solution to this problem. Therefore, we have had to resort
to other procedures to obtain a conformation which would
lie in the potential energy well containing the minimum
corresponding to that of the native protein. If such

7

a conformation were attained, presently-available minimization procedures would lead to the minimum.

V. Dominance of Short-Range Interactions

For reasons outlined in section VII, one of the
approaches for finding alternative methods led to a
consideration of the possible dominance of short-range
interactions (33). This investigation led to the concept (33) that the conformation of an amino acid residue in a polypeptide or protein is determined in very
large measure, though not exclusively, by the short-
range interactions between a side chain and the atoms
of the backbone of the *same* amino acid residue, and is,
again in first approximation, essentially independent
of interactions with neighboring side chains or backbone portions of the chain. This view has recently
received further support from a statistical analysis
of the conformations of amino acid residues in globular
proteins by Finkelstein and Ptitsyn (14). Therefore,
let us trace the development and application of this
concept, in order to see how it may help us overcome
the multiple-minima problem.

VI. Definitions

As used here, the term "short-range" refers to an
interaction between the side chain of an amino acid
residue with its own backbone. The interaction between
the atoms of a given residue with those of any other
residue, nearby in the chain or more remote along the
chain (even though, possibly, nearby in space) is
termed "long-range."

VII. The θ-point

The treatment of an ideal homopolymer chain by
random-flight statistics leads to the conclusion that
some average linear dimension of the chain, *e.g.*, the
root-mean-square end-to-end distance $\langle r^2 \rangle^{1/2}$, varies
with the square-root of the molecular weight (15).

8

While long-range and excluded volume effects (not included in the random-flight calculation) tend to increase $<\bar{r}^2>^{1/2}$ beyond its *ideal* value, the choice of an appropriate (poor) solvent (in which polymer-polymer contacts are favored over polymer-solvent contacts) can reduce $<\bar{r}^2>^{1/2}$ to its *ideal* value (15). Under these conditions (*i.e.*, at the θ-point), the polymer-polymer and polymer-solvent interactions compensate the long-range and excluded volume effects, and the ideal value of $<\bar{r}^2>^{1/2}$ which results is determined entirely by short-range interactions (15). Although a protein in aqueous salt solution may not be at the θ-point, the possibility existed that its conformation, while not determined exclusively by short-range interactions, might nevertheless be dominated by them. As I will show here, the dominance of short-range interactions has been demonstrated for the formation of α-helical and non-helical portions of proteins (14,33), and for the formation of β-turns (39), and also extended structures (9).

VIII. Conformational Preferences Within a Single Peptide Unit

To examine the validity of the hypothesis that short-range interactions are dominant, a study was made (33) initially of the role of these interactions in α-helix formation for proteins of known structure. In particular, calculations were carried out to obtain the energy of interaction (in dipeptide units) of individual side chains in lysozyme with side chains that are nearest neighbors along the backbone, as well as with the backbone groups themselves. It was found that, for various initial backbone conformations (*viz.*, the right- and left-handed α-helices, α_R and α_L, respectively, and the antiparallel pleated sheet structure, β), the conformation of lowest energy after minimization was the same in most cases for a given amino acid residue and was independent of the nature of the next amino acid in the chain. Furthermore, the backbone structures corresponding to the lowest energy (*i.e.*,

α_R, β, or α_L) showed a high degree of correlation with the so-called helix-making or helix-breaking character of a residue, as determined by earlier *empirical* studies on the identification of α-helical regions in proteins (23,24,55,64). In other words, it appears that the short-range interactions within a given peptide unit may be the physical origin of the so-called helical potential of a residue. In addition, since the side chain-side chain interaction does not play a major role in determining conformation in most cases, the cooperativity among residues, which is necessary for the formation of a helical segment, may simply be the additive effect of placing some sequence of helix-making residues in a particular region. This suggested a model for helix formation in which each type of peptide unit in proteins of known amino acid sequence was assigned a designation h or c (helix-making or helix breaking, respectively), based on a study of the energy surface of the peptide unit. Then, from an examination of the h or c assignments for lysozyme, myoglobin, α-chymotrypsin and ribonuclease, empirical rules were formulated to distinguish between helical and nonhelical regions. These rules are: (a) an α-helical segment will be nucleated when at least four h residues in a row appear in the amino acid sequence and (b) this helical segment will continue growing toward the C-terminus of the protein until two c residues in a row occur, a condition that terminates the helical segment. With these rules, it was possible to predict the helical or nonhelical state of 78% of the residues of the four proteins mentioned above (33).

With the later availability of the X-ray structures of seven proteins, the validity of these rules was examined further (34). It was observed that, if a nonhelical dipeptide ever occurred at the C-terminus of a helical region, it had a low probability of occurring elsewhere in a helical region and as high as a 90% probability of occurring elsewhere in nonhelical regions; *i.e.*, two c residues in a row prevent further growth of a helical segment. It was also found that those residues designated as c's tended to predominate at the C-termini of helical segments. These results constitute

10

an experimental demonstration of the validity of rule
(b) above. Finkelstein and Ptitsyn (14) also made a
statistical analysis of the conformations of amino acid
residues in proteins of known structure, and came to
similar conclusions, viz., that short-range interactions
are dominant, in that single residues can be classified
as helix-making or helix-breaking and that side chain-
side chain interactions play a minor role in determining
the conformational preference of a given amino acid
residue.

At this point, it is of interest to consider the
factors which determine the conformational preference
of a given amino acid residue. The conformational entropy of a residue in the random coil state must be
overcome by favorable energetic factors in order for the
residue to be helix-making; otherwise, it will be helix-
breaking. Glycyl residues, with no side chains, have no
favorable energetic factors to enhance helix formation;
thus, the entropy of the coil makes glycyl residues
helix breaking (19). When a β-CH$_2$ group is added, the
resulting nonbonded interactions tend to favor the α_R
conformation (19,33). Thus, alanine is a helix-making
residue (19). While all amino acids besides glycine
have a β-CH$_2$ group, they are not all helix-making because of interactions involving groups beyond the β-
carbon; e.g., in Asn which is helix-breaking, electrostatic interaction between the polar side-chain group
and the polar backbone amide group de-stabilizes the
α_R conformation relative to other conformations. In
Gln and Glu, the electrostatic effect is weaker because
of the greater distance between the backbone amide group
and the polar side-chain group (resulting from entropic-
ally-favored extended side-chain conformations); hence,
the preferred conformation for Gln and Glu is α_R. Recently, an extensive series of conformational energy
calculations (including the computation of statistical
weights) was carried out for the N-acetyl-N'-methyl
amides of all twenty naturally-occurring amino acids
(38). From these calculations, it is possible to assess
how the various energetic factors contribute to the conformational preferences of each residue. For example,

the side chains of both Ser and Asp can form hydrogen
bonds with the nearby backbone amide groups when these
residues are in nonhelical conformations, as illustrated
in Figs. 4 and 5; thus, Ser and Asp are helix-breaking.

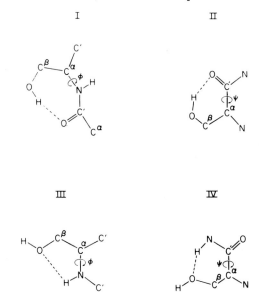

Fig. 4. Illustration of the types of hydrogen bonds
between serine side chains and the backbone
(38).

The above discussion suggests that, even though
Asp and Glu both contain side-chain carboxyl groups,
they cannot always be substituted one for another in
homologous proteins, especially if the residue under
consideration is next to a helix-breaking residue. In
such a case, a substitution of Asp for Glu in a *helical*
region would place two helix-breaking residues in a row,
and break the helix.

IX. Quantitative Specification of Helix-Making
and Helix-Breaking Character

Having demonstrated that the conformation of an
amino acid residue in a protein is determined largely

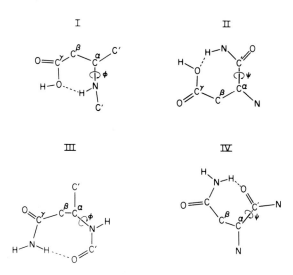

Fig. 5.   Illustration of the types of hydrogen bonds between aspartic acid and asparagine side chains and the backbone (38).

by short-range interactions, and, thus, in first approximation is essentially independent of the chemical nature of its neighbors, it becomes desirable to have a quantitative scale to specify the helix-making and helix-breaking character of the twenty naturally-occurring amino acids--instead of the earlier (33) assignment of all amino acids to two categories, h or c. A model which suggests itself is the helix-coil transition in homopolymers; i.e., the Zimm-Bragg nucleation and growth parameters σ and s (79), which characterize the transition curve, would appear to provide a quantitative basis for specifying the helix-making and breaking tendency of any amino acid in its corresponding homopolymer and, therefore, in a protein, since short-range interactions dominate in both cases. While σ and s can be computed from conformational energy calculations on homopolymers of amino acids (19), it is desirable to obtain these quantities directly from experiment. Because of certain experimental problems discussed elsewhere (62,71,72), homopolymers cannot be used for this purpose, and we have resorted instead to the experimental use of random

copolymers of two components--a helical host, for which
σ and s are known, and the guest residues; from the
effect of increasing amounts of the guest residues on
the helix-coil transition curve of the homopolymer of
the host residues, it is possible to determine σ and s
for the guest residues (62,71,72). An example of the
effect of alanine (as the guest residue) on the melting
behavior of a host polymer made of hydroxypropylgluta-
mine residues (48) is shown in Fig. 6. Thus far, these

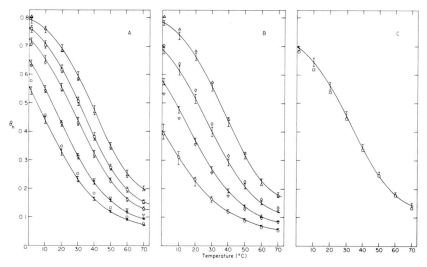

Fig. 6. Calculated melting curves for poly(hydroxypropylglutamine-co-L-alanine) copolymers in
water, together with experimental data (48).
The chain lengths and alanine contents of the
copolymers are:

A.   o, 422,14.1%   B.   o, 165,11.0%
     ∇, 880,20.6%   ∇, 322,19.5%
     ◊, 1010,32.0%   ◊, 536,30.4%
     □, 1413,38.3%   △, 1322,44.1%
     △, 1102,49.1%   C.   □, 365,38.1%

experiments have been carried out for the following
guest residues: Gly (3), Ala (48), Ser (30), Leu (2),
Phe (70) and Val (1), and the results are shown in
Fig. 7. It can be seen, e.g., that Gly and Ser are
helix breakers, Gly more so than Ser (because S < 1),

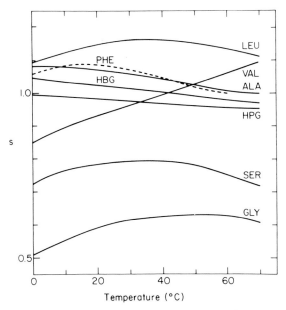

Fig. 7.   Temperature dependence of s (Zimm–Bragg growth
parameter) for various amino acid residues in
water (1).

and Ala and Leu are helix makers, Leu more so than Ala.
Since the experiments were carried out in aqueous solution (1-3, 30,48,70,71), the resulting experimental
values of σ and s contain all energetic and entropic
contributions (including solvation) which determine the
conformational preference of the residue.

X.   Helix Probability Profiles

The experimental value of σ and s can be used to
obtain information about the conformation of any specific
sequence of amino acids, e.g., that of a protein.   However, since the values of σ and s were obtained from the
Zimm–Bragg theory, which is based on the one-dimensional
Ising model, we cannot treat the *native* protein molecule
since its conformation is, in some measure, influenced
by long-range interactions which are not taken into
account in the Zimm–Bragg theory.   Since the *denatured*

protein is devoid of tertiary structure and hence, presumably, of long-range interactions other than excluded
volume effects, the polypeptide conforms to the one-
dimensional Ising model. Thus, above the denaturation
temperature, we may apply the Zimm-Bragg formulation to
this copolymer of ~ 20 amino acids to determine the
probability that any given residue of the chain will be
in the α_R or in the random coil conformation, respectively (36). I will then show that there is a correlation between the calculated α_R probability profile of
the *denatured* protein and the experimentally observed
α_R regions in the corresponding *native* structures; *i.e.*,
in many cases, those regions in the denatured protein
which exhibit a propensity for being in the α_R conformation correspond to the α_R regions observed in the
native protein.

The partition function Z, and the probability,
$P_H(i)$, that the ith amino acid (of type A) in a chain
of N residues is in the α_R conformation are given by

$$Z = (0,1) \left[\prod_{j=1}^{N} \underset{\sim}{W}_A(j) \right] \begin{pmatrix} 1 \\ 1 \end{pmatrix} \tag{1}$$

and

$$P_H(i) = (0,1) \left[\prod_{j=1}^{i=1} W_A(j) \right] \frac{\partial W_A(i)}{\partial \ln s_A(i)} \left[\prod_{j=i+1}^{N} W_A(j) \right] \begin{pmatrix} 1 \\ 1 \end{pmatrix} \Big/ Z \tag{2}$$

where $\underset{\sim}{W}_A(j)$ is the matrix of statistical weights for
the jth residue which is of amino acid type A, *viz.*

$$\underset{\sim}{W}_A(j) = \begin{pmatrix} s_A(j) & 1 \\ \sigma_A(j)s_A(j) & 1 \end{pmatrix} \tag{3}$$

$s_A(j)$ is the statistical weight assigned to this residue when it is in an α_R conformation and preceded by a residue in the α_R conformation, and $\sigma_A(j)s_A(j)$ is the statistical weight assigned to this residue when it is in an α_R conformation and preceded by a residue in the random coil conformation. The use of eq. 2 to compute $P_H(i)$ automatically includes the cooperativity which is characteristic of the nearest-neighbor one-dimensional Ising model.

The values of σ and s, determined in aqueous solution (as indicated in section IX), apply to the *initial* folding of a polypeptide chain; *i.e.*, the groups of a denatured protein are exposed to water and the values of σ and s (obtained from random copolymers in water) determine their *tendency* to form helices in the *denatured* protein, *before* the onset of globularity buries the helix in the non-aqueous interior of the protein.

Pending the acquisition of data, such as those of Fig. 7, for the remainder of the twenty naturally-occurring amino acids, the set of amino acids has been grouped into three categories (all with σ taken as 5×10^{-4}), *viz.*, helix breakers (with s = 0.385), helix formers (with s = 1.05), and helix indifferent (with s = 1.00). Taking into account the limited data of Fig. 7, the earlier h and c assignments of Kotelchuck and Scheraga (33), and the results of an information-theory analysis by Pain and Robson (46), the amino acids are assigned as in Table 1. It should be emphasized that these *tentative* values of σ and s are used here only pending completion of experiments which will fill out Fig. 7 for the remaining amino acids.

Helix probability profiles for 11 proteins have been calculated from eq. 2, using the values of σ and s discussed above and the assignments of Table 1[3] (40).

---

[3]More detailed information about the conformational state of each residue is provided by a recently-developed eight-state model for the helix-coil transition in homopolymers and specific-sequence copolymers (19,20).

TABLE 1

Assignment of Amino Acid Residues to Three Categories
According to Helix-Forming Power

Helix Breaker	Helix Indifferent	Helix Former
Gly	Lys	Val
Ser	Tyr	Gln
Pro	Asp	Ile
Asn	Thr	His
	Arg	Ala
	Cys	Trp
	Phe	Met
		Leu
		Glu

An example of some of these curves is shown in Fig. 8;

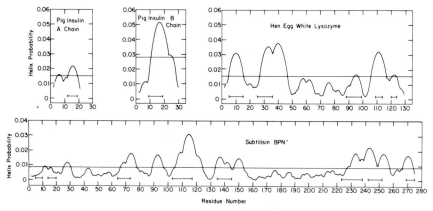

Fig. 8. Helix probability profiles for three proteins
in the denatured form (36). The ordinates
correspond to $P_H(i)$ computed from eq. 2. The
horizontal bars ($\longmapsto$) denote those regions of
the protein found to be in the α_R conformation
by x-ray diffraction analysis. The horizontal
line is the computed mean value of $P_H(i)$.

the occurrence of helix in the *native* protein is indicated by the short horizontal lines in each diagram.

From these curves it appears that there is a close correlation between the propensity of a particular amino acid residue to be in the α_R conformation in the denatured protein and its occurrence in a helical region in the globular structure of the corresponding native protein. On this basis, we suggested (36) that, during renaturation, the protein chain acquires *specific* long-range interactions which stabilize the helical regions which tend to form in certain portions of the chain; *i.e.*, folding of the polypeptide chain into the native conformation of a protein is thought to occur by incipient formation of α-helical or other ordered structural regions (among those residues with a propensity to be helical) stabilized by specific long-range interactions, with the remainder of the protein molecule then folding around these stabilized helical regions.

Consistent with this view, it is found (40) that, despite amino acid substitutions in a series of 27 species of cytochrome *c* proteins, there is a striking similarity in their helix probability profiles, and a good correlation with the location of the helical regions in the X-ray determined structure of the horse and bonito proteins (see Figs. 9 and 10). Among other things, the preponderance of *two* helix-breaking residues at the C-termini of helical sections (except at the C-terminus of the molecule where helix-breaking residues are not needed to break the helix), shown in Fig. 10, confirms the earlier, more primitive rule (*b*) mentioned in section VIII. It appears that amino acid substitutions may be tolerated in evolution, provided that the helix-making or helix-breaking tendency (*i.e.*, values of σ and s) of each amino acid residue is preserved, thereby enabling the altered protein to maintain the same three-dimensional conformation and, hence, the same biological function.

Application of this approach to lysozyme and α-lactalbumin (41), two different proteins with striking homologies in their amino acid sequence, led to very similar helix probability profiles (see Fig. 11). This result supports earlier suggestions (7,8,27) that the

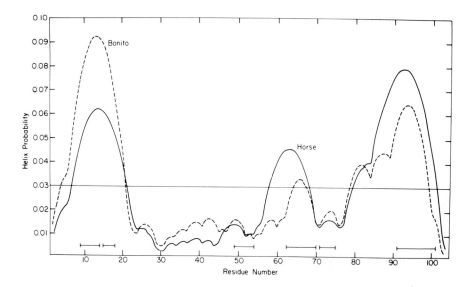

Fig. 9.   Helix probability profiles for horse and
bonito ferri-cytochrome c proteins (40).   The
ordinate corresponds to helix probability and
the abscissa to chain site; the horizontal
bars at the bottom denote those regions of
the native protein found to be in the α_R conformation by X-ray diffraction analysis.   The
horizontal line is the calculated value of
the overall mean helix content.

two proteins might have similar three-dimensional structures, and again demonstrates the conservative nature
of amino acid replacements, as far as helix-forming
power in homologous proteins is concerned, which was
found for the cytochrome c proteins.   In fact, on the
basis of the data of Fig. 11, we have used the X-ray
structure of lysozyme as a starting point for energy
minimization of the structure of α-lactalbumin (74)
(see section XI).

Species	N	5 10 15 20 25	50 55 60 65 70 75 80 85 90 95 100
Rabbit	104		
Whale	104		
Kangaroo	104		
Human	104		
Pig	104		
Horse	104		
Donkey	104		
Dog	104		
Lamprey eel	104		
Dogfish	104		
Tuna	103		
Bonito	103		
Chicken	104		
Penguin	104		
Pigeon	104		
Pekin duck	104		
Turtle	104		
Rattlesnake	104		
Bullfrog	104		
Screw worm fly	107		
Fruit fly	107		
Samia cynthia	107		
Tobacco moth	107		
Candida krusei	109		
Neurospora crassa	107		
Baker's yeast	108		
Wheat germ	112		
X-ray structure			

Fig. 10. Predicted α_R-helical segments for 27 cytochrome c proteins in regions between residues
4-25 and 47-104 (40). N is the total number
of amino acid residues in each single-chain
protein; referring to Table I, the helix
breakers are designated by •, the helix
formers by o and the helix-indifferent residues by blank spaces; the predicted helical
segments are indicated by square brackets;
the observed ones are given at the bottom of
the figure.

XI. Refinement of X-ray Structures, and Computation of Homologous Protein Structures

Having referred to X-ray structures and to α-
lactalbumin, let us digress at this point to consider
some interesting computations that can be carried out
even before we solve the multiple-minima problem.
First of all, it is possible to refine X-ray
structures of proteins (which are usually obtained at
~ 2Å-resolution) to atomic resolution by energy minimization. Since the 2Å-resolution structure is known,
we are already in the right potential energy well for
the energy minimization. We have developed a three-
stage procedure for this purpose (73,75,76), and have

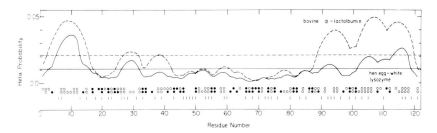

Fig. 11. Helix probability profiles for hen egg-white
          lysozyme (———) and bovine α-lactalbumin
          (----) (41). The ordinate corresponds to
          helix probability and the abscissa to chain
          site; the horizontal solid and dashed lines
          are the calculated values of the overall
          mean helix content for the complete chains of
          lysozyme and α-lactalbumin, respectively; the
          helix breakers are designated by ●, the helix
          formers by o and the helix-indifferent resi-
          dues by blank spaces; the short vertical lines
          indicate those residues that are identical in
          both proteins.

carried out these computations most extensively for
actinomycin D (52), rubredoxin (57) and lysozyme (75),
and less extensively for α-chymotrypsin (51). It is
necessary to have such energetically-refined protein
structures in order to obtain meaningful information
from calculations on enzyme-substrate interactions (see
section XVII).

     Secondly, if the three-dimensional structure of one
protein is known, then its conformation can be used as a
starting one for the computation of the structure of a
homologous protein. Here again, the multiple-minima
problem is essentially bypassed because of the availability of the structure of one of the homologous proteins
By this procedure, the structure of α-lactalbumin has
been computed (74) from that of lysozyme. Similar calculations are in progress (69) for the family of homologous serine proteases (chymotrypsin, trypsin, elastase
and thrombin).

XII. β-turns

    Having digressed in section XI, we now return to our main theme, *viz.*, that of trying to obtain a rough conformation (one in the right potential energy well) for subsequent energy minimization. Thus far, we have considered the possibility of predicting the regions of α_R-helical conformation, and their role in protein folding. Before considering the prediction of extended-structure regions, we must discuss a problem which arises because of the inherent instability of *isolated short* segments of α-helix. While the protein chain must fold in order to enable remote helical or other ordered structural regions to approach each other to be stabilized by long-range interactions, it is felt that these long-range interactions are not brought into play by a *chance* encounter of the ordered regions (39). In fact, it would probably take about 10^{40} years for such a chance encounter to occur if all of conformation space were explored. Instead, there is a tendency for bends or β-turns (see Fig. 12) to occur among certain amino acid residues (again, arising from short-range interactions), thereby "directing" the encounter of the

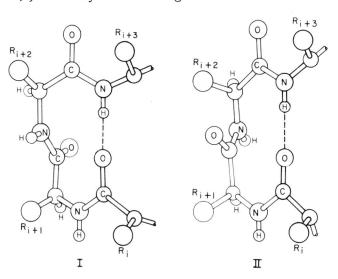

Fig. 12. Type I and II β bends.

ordered regions. From a statistical analysis of the
amino acid composition of the bends in three proteins,
it has been possible to formulate rules for the existence of β-turns in general. Application of these rules
to several other proteins led to a high degree of correlation between the predicted regions where the β-turns
should appear, and their existence in the X-ray determined structure (39). A discussion of β-turns in proteins has also been presented by Kuntz (35). It is of
interest that residues like Gly, Ser and Asp, which have
a low tendency toward helix formation, have a high propensity to form β-turns.

We have recently completed a study (37) of the
bends found in the native structures of eight proteins.
The 135 bends which were located could be grouped among
ten types, and over 40% of the bends did not possess a
hydrogen bond between the C=O of residue i and the NH
of residue i+3 (see Fig. 12). In addition, conformational energy calculations were carried out on three
pentapeptides with amino acid sequences found as bends
in the native structure of α-chymotrypsin. The results
indicate that the bends occur not only in the whole
molecule, but also in the pentapeptide; *i.e.*, the observed bends were the conformations of lowest energy
even in the pentapeptides. The stability of the bends,
compared to those of other structures, arises principally from side chain-backbone interactions (*e.g.*, a
hydrogen bond between the side-chain COO⁻ of Asp in
position i+3 and the backbone NH of the residue in position i) rather than from i to i+3 backbone-backbone
hydrogen bonds. This result is consistent with the
observation (39) that residues with small polar side
chains, such as Ser, Thr, Asp and Asn, are found frequently in bends, presumably because these residues can
interact most strongly with their immediate backbones.
Water may also play a role in stabilizing bends among
these polar residues, especially since such bends usually occur on the surface of a protein. Experiments are
in progress (63) to attempt to detect bends in short
oligopeptides in water.

XIII.   Extended Structures

Short-range interactions also endow certain amino acid residues with a propensity for adopting an extended-structure conformation; long runs of such extended structures can participate in the formation of parallel or anti-parallel β-structures. The conformational energy calculations on the N-acetyl N'-methyl amides of the twenty naturally occurring amino acids (38), cited in section III, provide a basis for assigning probabilities of occurrence of extended structures for each amino acid. The probabilities were multiplied together over a sequence of seven residues (to take medium-range as well as short-range interactions into account; see section XV) to compute the propensity of a whole segment of the polypeptide chain to form an extended structure. Examples of probability profiles for extended structures are shown in Fig. 13, from which it can be seen that

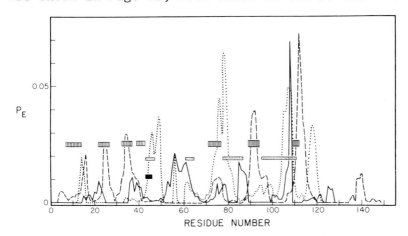

Fig. 13.   Probability profiles for extended structures in lysozyme (solid line), ribonuclease (dotted line), and staphylococcal nuclease (dashed line) (9). Horizontal bars indicate the observed position of extended structures in the three proteins: lysozyme (▬▬), ribonuclease (▭▭▭▭), and staphylococcal nuclease (ⅢⅢⅢ).

25

there is a strong positive correlation between the probability profiles and the experimentally occurring extended
structure. In an analogous way to the method used for
β-turns (see section XII), a statistical analysis of extended structures in proteins (with a correlation over
nine residues (see section XV)) has also led to rules for
the prediction of extended structures (9).

XIV. Nature of Folding

From the above discussion, the following picture
of the folding of the polypeptide chain emerges: helical or other ordered structural regions (such as extended structures) tend to form in certain regions of
the amino acid sequence of the polypeptide chain, in
response to short-range interactions. These are stabilized, however, only when long-range interactions come
into play. This is brought about by the formation of
β-turns among *specific* amino acid residues, also on the
basis of short-range interactions, thereby enabling the
ordered regions to approach each other. The remainder
of the polypeptide chain then folds around these one or
more regions of interacting ordered structures. Keep-
ing in mind our original aim of deducing rules for predicting conformational states of residues, and hence a
rough structure which would lie in the correct potential
energy well for subsequent energy minimization, we now
consider explicitly the role of longer-range interactions.

XV. Medium-Range Interactions

We know that an individual residue prefers a limited number of conformational states (38). In order to
see whether medium-, rather than longer-range, interactions might further limit this choice of conformational states, the nonamer segments, illustrated in
Fig. 14, were examined (53). The central (i^{th}) residue
was considered, successively, as part of a trimer,
pentamer, heptamer and nonamer for specific sequences
of amino acids from lysozyme; in fact the lysozyme molecule was divided into 14 such nonamer segments. In each

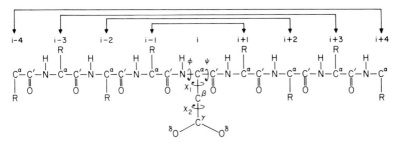

Fig. 14.   Definition of the central residue i in tri-,
           penta-, hepta-, and nonapeptides (53).   The
           dihedral angles ϕ, Ψ, χ_1, and χ_2 of the cen-
           tral residue are indicated for Asp, as an
           example.

oligomer, all residues but the ith were kept in their
X-ray determined structure, and the energy of the whole
oligomer was minimized by starting from five different
initial conformations of the ith residue.   As the length
of the oligomer increased (up to nine residues), the
tendency for the central residue to adopt the observed
conformation (specified in terms of the dihedral angles
of the backbone) increased.   In fact, for 10 of the 14
nonamer segments of lysozyme, the lowest-energy conformation of the central residue was the observed one.

XVI.   Predictive Rules

With the information in section XV available, *viz.*,
that up to four residues on either side of a given residue influence the conformation of a given residue, pre-
dictive rules were developed (9) based on nonamer units;
i.e., the probabilities of occurrence of nine residues
in specific conformations are multiplied together to
determine the propensity of the central residue to be
in the α_R-helical, extended-structure, or β-turn, or
other ("coil") conformation.   While these rules do not
yet lead to 100% validity of predictions, they provide
initial information about the distribution of each residue of a polypeptide chain among a discrete number (*viz.*,
four) of conformational states.

It is worth mentioning that the empirical probabilities (which are assigned to each residue, and then multiplied together in nonamer segments) were determined from the known structures of eight proteins; thus, in contrast to probabilities obtained from conformational energies on *single* residues (38), they reflect the long-range interactions as well, of course, as the short-range interactions present in proteins.

Focusing attention on nonamer segments (see, *e.g.*, Fig. 14) we now have a basis for assigning one of four states to each residue in a nonamer in a protein. The energy of the whole nonamer is then minimized with respect to the dihedral angles of only the central residue. Then an overlapping nonamer is selected (shifted by one residue toward the C-terminus of the chain), and the four-state model is again applied to all residues except the one whose dihedral angles were varied in the previous step; for this residue, the minimum-energy *dihedral angles* are assigned. This process is illustrated schematically in Fig. 15. Thus, by moving down the whole length of a protein chain (and possibly repeating the process several times) we may be able to see whether we are approaching the native conformation by successive approximation. Once the structure begins to resemble that of the native protein, we would be in the correct potential energy well to start the minimization of the energy of the whole protein, not just that of nonamer segments. This procedure is now being tested, but no results are as yet available.

Much effort is being devoted in many laboratories to try to deduce empirical rules for predicting the conformations of individual residues in proteins. My view is that these predictions will not provide a three-dimensional picture of a protein but, at best, along with information about longer-range interactions, might provide a rough structure (in the correct potential energy well) from which subsequent energy minimization will lead to the native structure of a protein.

28

Step	1	2	3	4	5	6	7	8	9	10	11	12	13 ....
1	β	β	α	α	α	α	α	β	β				
2	β	β	α	α	$\phi=-52°$ $\psi=-48°$	α	α	β	β				
3		β	α	α	$\phi=-52°$ $\psi=-48°$	α	α	β	β	β			
4		β	α	α	$\phi=-52°$ $\psi=-48°$	$\phi=-51°$ $\psi=-49°$	α	β	β	β			
5			α	α	$\phi=-52°$ $\psi=-48°$	$\phi=-51°$ $\psi=-49°$	α	β	β	β	β		
6			α	α	$\phi=-52°$ $\psi=-48°$	$\phi=-51°$ $\psi=-49°$	$\phi=-52°$ $\psi=-49°$	β	β	β	β		

etc.

Fig. 15. Schematic illustration of successive steps in the computation of the dihedral angles of a protein. The amino acid sequence is specified by the numbers 1, 2, ... 13 ...., and the backbone conformations of each residue are specified *initially* as α, β, etc., but in terms of the dihedral angles ϕ and ψ after each computational step. In step 1, one of four conformational states is assigned to each residue in a nonapeptide. In step 2, the energy of the nonapeptide is minimized with respect to the dihedral angles of residue 5, etc.

XVII. Enzyme-Substrate Interactions

Before closing, it might be of interest to ask what conformational questions will interest us, if we are able to predict the three-dimensional structure for a given amino acid sequence. One important application is the use of energy-minimized protein structures to study enzyme-substrate interactions. We have actually made a start on this problem (49) by treating the bimolecular complex of enzyme and substrate. Subsequently, we have extended these calculations (50) to allow both the enzyme and substrate to change conformation as the non-covalent Michaelis complex is formed. Examples of computed low-energy structures of complexes of α-chymotrypsin with single-residue and tripeptide substrates are shown in Figs. 16 and 17, respectively.

29

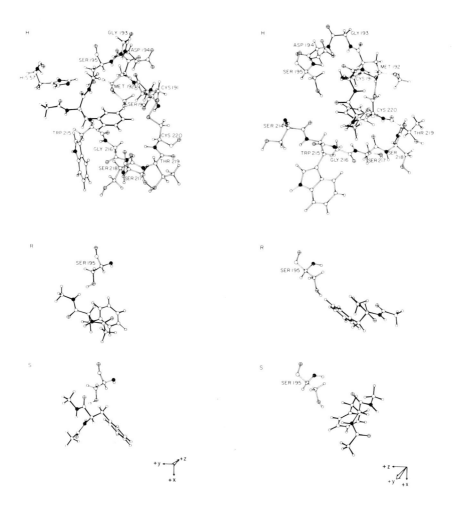

Fig. 16.   The favorable dispositions (H, R and S) of
           different conformations of N–acetyl N'–
           methyl L-phenylalanine amide in the active
           site cleft of α–chymotrypsin (50).

Also, calculated low-energy structures of the covalent
acyl intermediate are shown in Fig. 18.   Such calculations should serve to help interpret the wide range of
rates of acylation of α–chymotrypsin observed (6) when
the phenylalanine residue is flanked by different amino

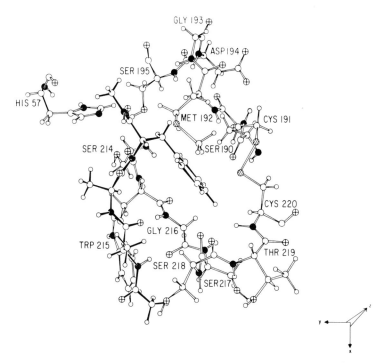

Fig. 17.   A low-energy conformation of the substrate
N-acetyl N'-methyl L-alanyl-L-alanyl-L-
phenylalanine amide in the active site cleft
of α-chymotrypsin. The substrate is shown
with solid-line bonds, while the enzyme has
open-line bonds (50).

acids which contribute to the binding energy, and also
to determine the orientation of the peptide bond to be
hydrolyzed with respect to the side chain of Ser 195,
which is acylated.

XVIII.   Summary

While we have not yet been able to predict the
three-dimensional structure of a protein from its amino
acid sequence, because of the multiple-minima problem,
we have learned a great deal about the factors which
influence the local conformations of amino acid residues in a chain. Hopefully, this information can be

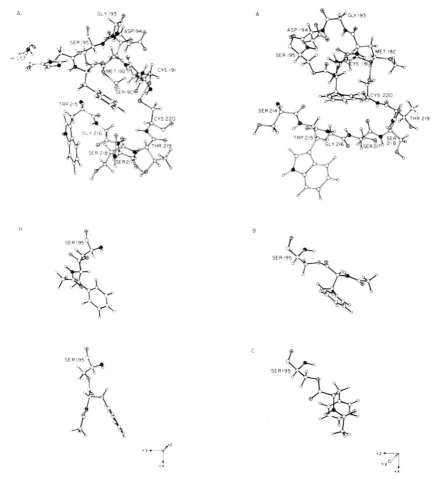

Fig. 18. The favorable conformations (A, B and C) of
          the acyl intermediate of N-acetyl-L-phenyla-
          lanine amide and α-chymotrypsin (50).

used, in the manner indicated in section XVI, or in
some other way, to surmount the multiple-minima problem.
Even before this problem is solved, we have already
been able to refine the structures of proteins to
atomic resolution (starting with 2Å resolution X-ray
structures) and to compute the structure of a protein

from the known structure of a homologous one. Finally, these same techniques can now be used to compute the preferred conformations of enzyme-substrate complexes and of their covalent intermediates.

References

1. Alter, J., Andreatta, R.H., Taylor, G.T. and Scheraga, H.A. Helix-coil stability constants for the naturally occurring amino acids in water. VIII. Valine parameters from random poly(hydroxy-propylglutamine-co-L-valine) and Poly(hydroxy-butylglutamine-co-L-valine). Macromolecules <u>6</u>, 564 (1973).

2. Alter, J., Taylor, G.T. and Scheraga, H.A. Helix-coil stability constants for the naturally occur-ring amino acids in water. VI. Leucine para-meters from random poly(hydroxypropylglutamine-co-L-leucine and poly(hydroxybutylglutamine-co-L-leucine). Macromolecules <u>5</u>, 739 (1972).

3. Ananthanarayanan, V.S., Andreatta, R.H., Poland, D. and Scheraga, H.A. Helix-coil stability constants for the naturally occurring amino acids in water. III. Glycine parameters from random poly(hydroxy-butylglutamine-co-glycine). Macromolecules <u>4</u>, 417 (1971).

4. Anfinsen, C.B. On the possibility of predicting tertiary structure from primary sequence. In New Perspectives in Biology, ed. M. Sela, Elsevier, Amsterdam, p. 42, 1964.

5. Barnard, E.A. and Stein, W.D. The histidine resi-due in the active centre of ribonuclease. J. Mol. Biol. <u>1</u>, 339, 350 (1959).

6. Baumann, W.K., Bizzozero, S.A. and Dutler, H. Kinetic investigation of the α-chymotrypsin-catalyzed hydrolysis of peptide substrates. The

relationship between peptide structure N-terminal
to the cleaved bond and reactivity, in press
(private communication from Dr. Dutler).

7. Brew, K., Vanaman, T.C. and Hill, R.L. Comparison of the amino acid sequence of bovine α-lactalbumin and hen's egg white lysozyme. J. Biol. Chem.
242, 3747 (1967).

8. Browne, W.J., North, A.C.T., Phillips, D.C., Brew,
K., Vanaman, T.C. and Hill, R.L. A possible three-
dimensional structure of bovine α-lactalbumin
based on that of hen's egg-white lysozyme. J. Mol.
Biol. 42, 65 (1969).

9. Burgess, A.W., Ponnuswamy, P.K. and Scheraga, H.A.
Analysis of conformations of amino acid residues
and prediction of backbone topography in proteins.
Israel J. Chem. 12, 2177 (1973).

10. Burgess, A.W. and Scheraga, H.A. Stable conformations of dipeptides. Biopolymers, in press.

11. Crippen, G.M. and Scheraga, H.A. Minimization of
polypeptide energy VIII. Application of the de-
flation technique to a dipeptide. Proc. Nat. Acad.
Sci. USA 64, 42 (1969).

12. Crippen, G.M. and Scheraga, H.A. Minimization of
polypeptide energy. XI. The method of gentlest
ascent. Arch. Biochem. Biophys. 144, 453, 462
(1971).

13. Crippen, G.M. and Scheraga, H.A. Minimization of
polypeptide energy. XII. The methods of partial
energies and cubic subdivision. J. Computational
Phys. 12, 491 (1973).

14. Finkelstein, A.V. and Ptitsyn, O.B. Statistical
analysis of the correlation among amino acid residues in helical, β-structural and non-regular

regions of globular proteins. J. Mol. Biol. 62, 613 (1971).

15. Flory, P.J., _Principles of Polymer Chemistry_, Cornell University Press, Ithaca, N.Y. (1953), chap. 14.

16. Gibson, K.G. and Scheraga, H.A. Minimization of polypeptide energy. I. Preliminary structures of bovine pancreatic ribonuclease S-peptide. Proc. Nat. Acad. Sci. USA 58, 420 (1967).

17. Gibson, K.G. and Scheraga, H.A. Minimization of polypeptide energy VII. Second derivatives and statistical weight of energy minima for deca-L-alanine. Proc. Nat. Acad. Sci. USA 63, 9, 242 (1969).

18. Gibson, K.G. and Scheraga, H.A. Minimization of polypeptide energy IX. A procedure for seeking the global minimum of functions with many minima. Computers and Biomedical Research 3, 375 (1970).

19. Gō, M., Gō, N. and Scheraga, H.A. Molecular theory of the helix-coil transition in polyamino acids. III. Evaluation and analysis of s and σ for polyglycine and poly-L-alanine in water. J. Chem. Phys. 54, 4489 (1971).

20. Gō, N., Lewis, P.N., Gō, M. and Scheraga, H.A. A model for the helix-coil transition in specific-sequence copolymers of amino acid. Macromolecules 4, 692 (1971).

21. Gross, E. and Witkop, B. Nonenzymatic cleavage of peptide bonds: the methionine residues in bovine pancreatic ribonuclease. J. Biol. Chem. 237, 1856 (1962).

22. Gundlach, H.G., Stein, W.H. and Moore, S. The nature of the amino acid residues involved in the

inactivation of ribonuclease by iodoacetate. J. Biol. Chem. 234, 1754, 1761 (1959).

23. Guzzo, A.V. The influence of amino-acid sequence on protein structure. Biophys. J. 5, 809 (1965).

24. Havsteen, B.H. A study of the correlation between the amino acid composition and the helical content of proteins. J. Theoret. Biol. 10, 1 (1966).

25. Heinrikson, L. On the alkylation of amino acid residues at the active site of ribonuclease. J. Biol. Chem. 241, 1393 (1966).

26. Heinrikson, L., Stein, W.H., Crestfield, A.M. and Moore, S. The reactivities of the histidine residues at the active site of ribonuclease toward halo acids of different structures. J. Biol. Chem. 240, 2921 (1965).

27. Hill, R.L., Brew, K., Vanaman, T.C., Trayer, I.P. and Mattock, P. The structure, function, and evolution of α-lactalbumin. Brookhaven Symp. Biol. 21, 139 (1968).

28. Hirs, C.H.W., Halmann, M. and Kycia, J.H. The reactivity of certain functional groups in ribonuclease A towards substitution by 1-fluoro-2,4-dinitrobenzene. Inactivation of the enzyme by substitution at the lysine residue in position 41. In Biological Structure and Function, ed. T.W. Goodwin and O. Lindberg, Academic Press, New York, 1961, Vol. 1, p. 41.

29. Hirs, C.H.W., Moore, S. and Stein, W.H. The sequence of the amino acid residues in performic acid-oxidized ribonuclease. J. Biol. Chem. 235, 633 (1960).

30. Hughes, L.J., Andreatta, R.H. and Scheraga, H.A. Helix-coil stability constants for the naturally occurring amino acids in water. V. Serine parameters from random poly(hydroxybutylglutamine-co-L-serine). Macromolecules 5, 187 (1972).

31. Kartha, G., Bello, J. and Harker, D. Tertiary structure of ribonuclease. Nature 213, 862 (1967).

32. Kendrew, J.C., Dickerson, R.E., Strandberg, B.E. Hart, R.G. and Davies, D.R. Structure of myoglobin. Nature 185, 422 (1960).

33. Kotelchuck, D. and Scheraga, H.A. The influence of short-range interactions on protein conformation. I. Side chain-backbone interactions within a single peptide unit. Proc. Nat. Acad. Sci. USA 61, 1163 (1968); II. A model for predicting the α-helical regions of proteins. Proc. Nat. Acad. Sci. USA 62, 14 (1969).

34. Kotelchuck, D., Dygert, M. and Scheraga, H.A. The influence of short-range interactions on protein conformation. III. Dipeptide distributions in proteins of known sequence and structure. Proc. Nat. Acad. Sci. USA 63, 615 (1969).

35. Kuntz, I.D. Protein folding. J. Am. Chem. Soc. 94, 4009 (1972).

36. Lewis, P.N., Gō, N., Gō, M, Kotelchuck, D. and Scheraga, H.A. Helix probability profiles of denatured proteins and their correlation with native structures. Proc. Nat. Acad. Sci. USA 65, 810 (1970).

37. Lewis, P.N., Momany, F.A. and Scheraga, H.A. Chain reversals in proteins. Biochim. et Biophys. Acta 303, 211 (1973).

38. Lewis, P.N., Momany, F.A. and Scheraga, H.A.
Energy parameters in polypeptides. VI. Conformational energy analysis of the N-acetyl N'-methyl
amides of the twenty naturally occurring amino
acids. Israel J. Chem. 11, 121 (1973).

39. Lewis, P.N., Momany, F.A. and Scheraga, H.A.
Folding of polypeptide chains in proteins: a
proposed mechanism for folding. Proc. Nat. Acad.
Sci. USA 68, 2293 (1971).

40. Lewis, P.N. and Scheraga, H.A. Predictions of
structural homologies in cytochrome c proteins.
Arch. Biochem. Biophys. 144, 576 (1971).

41. Lewis, P.N. and Scheraga, H.A. Prediction of
structural homology between bovine α-lactalbumin
and hen egg white lysozyme. Arch. Biochem. Biophys. 144, 584 (1971).

42. Momany, F.A., Carruthers, L.M., McGuire, R.F. and
Scheraga, H.A. Intermolecular potentials from
crystal data. J. Phys. Chem., submitted.

43. Momany, F.A., Goldman, A., Hesselink, F. Th. and
Scheraga, H.A. Prediction of the sense of the
α-helix in several homopolyamino acids. in preparation.

44. Nemethy, G. and Scheraga, H.A. Theoretical determination of sterically allowed conformations
of a polypeptide chain by a computer method.
Biopolymers 3, 155 (1965).

45. Niu, G.C.C., Gō, N. and Scheraga, H.A. Calculation of the conformation of the pentapeptide
cyclo(glycylglycylglycylprolylprolyl). III.
Treatment of a flexible molecule. Macromolecules
6, 91 (1973).

46. Pain, R.H. and Robson, B. Analysis of the code relating sequence to secondary structure in proteins. Nature 227, 62 (1970).

47. Perutz, M.F., Rossman, M.G., Cullis, A.F., Muirhead, H., Will, G., North, A.C.T., Phillips, D.C. and Shore, V.C. Structure of haemoglobin. Nature 185, 416 (1960).

48. Platzer, K.E.B., Ananthanarayanan, V.S., Andreatta, R.H. and Scheraga, H.A. Helix-coil stability constants for the naturally occurring amino acids in water. IV. Alanine parameters from random poly-(hydroxypropylglutamine-co-L-alanine). Macromolecules 5, 177 (1972).

49. Platzer, K.E.B., Momany, F.A. and Scheraga, H.A. Conformational energy calculations of enzyme-substrate interactions. II. Computation of the binding energy for substrates in the active site of α-chymotrypsin. Intntl. J. Peptide and Protein Research 4, 201 (1972).

50. Platzer, K.E.B. and Scheraga, H.A., unpublished results.

51. Platzer, K.E.B., Warme, P.K., Momany, F.A. and Scheraga, H.A., work in progress.

52. Ponnuswamy, P.K., McGuire, R.F. and Scheraga, H.A. Refinement of the molecular structure of actinomycin D by energy minimization. Intntl. J. Peptide Protein Res. 5, 73 (1973).

53. Ponnuswamy, P.K., Warme, P.K. and Scheraga, H.A. Role of medium-range interactions in proteins. Proc. Nat. Acad. Sci. USA 70, 830 (1973).

54. Potts, J.T., Berger, A., Cooke, J. and Anfinsen, C.B. A reinvestigation of the sequence of residues 11 to 18 in bovine pancreatic ribonuclease. J. Biol. Chem. 237, 1851 (1962).

55. Prothero, J.W. Correlation between the distribution of amino acids and alpha helices. Biophys. J. 6, 367 (1966).

56. Ramachandran, G.N. and Sasisekharan, V. Conformation of polypeptides and proteins. Adv. Protein Chem. 23, 283 (1968).

57. Rasse, D., Warme, P.K. and Scheraga, H.A. Refinement of the structure of rubredoxin. Biochemistry, to be submitted.

58. Ryle, A.P., Sanger, F., Smith, L.F. and Kitai, R. The disulphide bonds of insulin. Biochem. J. 60, 541 (1955).

59. Scheraga, H.A. Calculations of conformations of polypeptides. Adv. Phys. Org. Chem. 6, 103 (1968).

60. Scheraga, H.A., _Protein Structure_, Academic Press, New York, 1961.

61. Scheraga, H.A. Structural studies of pancreatic ribonuclease. Fed. Proc. 26, 1380 (1967).

62. Scheraga, H.A. Theoretical and experimental studies of conformations of polypeptides. Chem. Revs. 71, 195 (1971).

63. Scheraga, H.A., Lewis, P.N., Momany, F.A., Von Dreele, P.H., Burgess, A.W. and Howard, J.C. Hairpin bends in oligopeptides and proteins. Fed. Proc. 32, 495 (1973).

64. Schiffer, M. and Edmundson, A.B. Use of helical wheels to represent the structures of proteins and to identify segments with helical potential. Biophys. J. 7, 121 (1967).

65. Silverman, D.N. and Scheraga, H.A. Hairpin bend and interhelical interactions in α-helical poly-(L-alanine) in water. Arch. Biochem. and Biophys. 153, 449 (1972).

66. Smyth, D.G., Stein, W.H. and Moore, S. On the sequence of residues 11 to 18 in bovine pancreatic ribonuclease. J. Biol. Chem. 237, 1845 (1962).

67. Smyth, D.G., Stein, W.H. and Moore, S. The sequence of amino acid residues in bovine pancreatic ribonuclease: revisions and confirmations. J. Biol. Chem. 238, 227 (1963).

68. Spackman, D.H., Stein, W.H. and Moore, S. The disulfide bonds of ribonuclease. J. Biol. Chem. 235, 648 (1960).

69. Swenson, M.K. and Scheraga, H.A., work in progress.

70. Van Wart, H.E., Taylor, G.T. and Scheraga, H.A. Helix-coil stability constants for the naturally occurring amino acids in water. VII. Phenylalanine parameters from random poly(hydroxypropyl-glutamine-co-L-phenylalanine). Macromolecules 6, 266 (1973).

71. Von Dreele, P.H., Lotan, N., Ananthanarayanan, V.S., Andreatta, R.H., Poland, D. and Scheraga, H.A. Helix-coil stability constants for the naturally occurring amino acids in water. II. Characterization of the host polymers and application of the host-guest technique to random poly(hydroxypropyl-glutamine-co-hydroxybutylglutamine). Macromolecules 4, 408 (1971).

72. Von Dreele, P.H., Poland, D. and Scheraga, H.A. Helix-coil stability constants for the naturally occurring amino acids in water. I. Properties of copolymers and approximate theories. Macromolecules 4, 396 (1971).

73. Warme, P.K., Gō, N. and Scheraga, H.A. Refinement of X-ray data on proteins. I. Adjustment of atomic coordinates to conform to a specified geometry. J. Computational Physics 9, 303 (1972).

74. Warme, P.K., Momany, F.A., Rumball, S.S., Tuttle, R.F. and Scheraga, H.A. Computation of structures of homologous proteins; α-lactalbumin from lysozyme. Biochemistry, in press.

75. Warme, P.K. and Scheraga, H.A. Refinement of the X-ray structure of lysozyme by complete energy minimization. Biochemistry, in press.

76. Warme, P.K. and Scheraga, H.A. Refinement of X-ray data on proteins. II. Adjustment of specified geometry to relieve atomic overlaps. J. Computational Physics 12, 49 (1973).

77. Warshel, A., Levitt, M. and Lifson, S. Consistent force field for calculation of vibrational spectra and conformations of some amides and lactam rings. J. Mol. Spectroscopy 33, 84 (1970).

78. Wyckoff, H.W., Tsernoglou, D., Hanson, A.W., Knox, J.R., Lee, B. and Richards, F.M. The three-dimensional structure of riobnuclease-S. J. Biol. Chem. 245, 305 (1970).

79. Zimm, B.H. and Bragg, J.K. Theory of the phase transition between helix and random coil in polypeptide chains. J. Chem. Phys. 31, 526 (1959).

STUDIES ON THE STRUCTURE OF GLUTAMIC DEHYDROGENASE

Henryk Eisenberg

Fogarty International Center
National Institutes of Health
Bethesda, Maryland
and
Department of Polymer Research[*]
The Weizmann Institute of Science
Rehovot, Israel

I am glad to be here to summarize a research project we have been working on in recent years - the relation between structure and function of glutamic dehydrogenase (GDH). I would like to acknowledge my
deep indebtedness to Gordon Tomkins who in 1966, during my stay at the National Institutes of Health, introduced me to the puzzling aspects of this problem
and infected me with the bug of curiosity which has
become a steady companion ever since. I would like to
apologize about the title, because nowadays when one
says "structure of an enzyme," actually a great deal
more is meant and one is expected to speak of the details of molecular structure at the atomic level. For-
tunately for us, when this work was started, and even
to the present day, the manufacture of crystals of glutamic dehydrogenase of quality suitable for the newer
tools of molecular torture had not been achieved. This
lecture will hopefully show that by classical light
scattering, sedimentation, viscosity, and related
studies it is possible to unravel some of the intri-
cacies of structure and the relation of structure to
function of this complex regulatory enzyme. Fortu-
nately the sequence of the enzyme is now known (25)

---

[*]Address for correspondence.

and possibly it will be possible in the near future to obtain good enough crystals for X-ray analysis.

Some of the slides I will show are probably not necessary for this audience but I will run over them quickly just in case someone is not familiar with the reaction catalyzed by glutamic dehydrogenase. I will first review some of the earlier work we have done and in the second part of my talk, discuss some work about to be published on an active crosslinked form of the enzyme which we have recently been able to produce and which is quite interesting in relation to structural and functional properties. First, I would like to say that much of this work was done by a former student of mine, Emil Reisler, who is now doing post-doctoral studies with William Harrington at Johns Hopkins University in Baltimore. All the electron microscopy (except for one slide due to the late Robin Valentine) is due to Robert Josephs, who also worked with Harrington and who is now in our laboratory.

Self Assembly of the Enzyme

Figure 1, taken from the paper of Olson and Anfinsen in 1952 (18), depicts the concentration dependence of the sedimentation coefficient of the enzyme

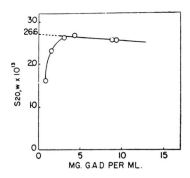

Fig. 1. Sedimentation coefficients of glutamic dehydrogenase in 0.05 M potassium phosphate buffer, pH 7.4 (from Reference 18).

dissolved in phosphate buffer, below 5 mg/ml concentration of protein. A sharp decrease in sedimentation coefficient was observed as the concentration was decreased. This puzzling phenomenon was, I believe, correctly diagnosed by the authors, and we know now that what is involved is a concentration-dependent, fully reversible dissociation of the enzyme. The sedimentation coefficient decreases because of disaggregation and the limiting sedimentation coefficient, $S_{20,w}^0$ (in the limit of vanishing enzyme concentration), is a little below 12 Svedberg units. These early measurements were thus essentially correct and pointed to some unusual properties of this enzyme system.

The reaction catalyzed by glutamic dehydrogenase, deamination of glutamic acid to yield α-ketoglutaric acid, is shown in Figure 2, and Figure 3, taken from

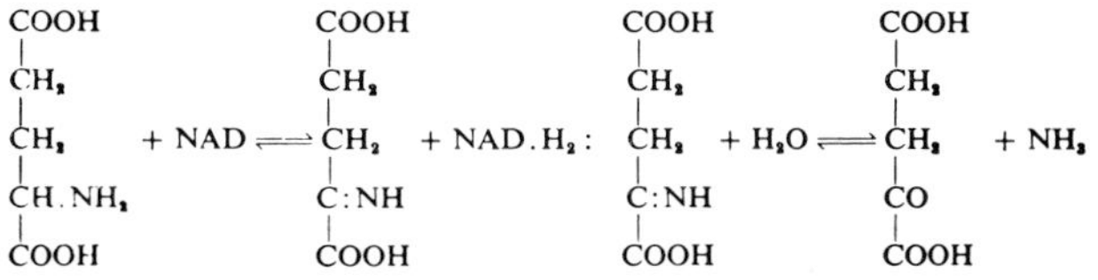

$$
\begin{array}{c}
\text{COOH} \\
| \\
\text{CH}_2 \\
| \\
\text{CH}_2 \\
| \\
\text{CH.NH}_2 \\
| \\
\text{COOH}
\end{array}
+ \text{NAD} \rightleftharpoons
\begin{array}{c}
\text{COOH} \\
| \\
\text{CH}_2 \\
| \\
\text{CH}_2 \\
| \\
\text{C:NH} \\
| \\
\text{COOH}
\end{array}
+ \text{NAD.H}_2 :
\begin{array}{c}
\text{COOH} \\
| \\
\text{CH}_2 \\
| \\
\text{CH}_2 \\
| \\
\text{C:NH} \\
| \\
\text{COOH}
\end{array}
+ \text{H}_2\text{O} \rightleftharpoons
\begin{array}{c}
\text{COOH} \\
| \\
\text{CH}_2 \\
| \\
\text{CH}_2 \\
| \\
\text{CO} \\
| \\
\text{COOH}
\end{array}
+ \text{NH}_3
$$

Fig. 2. Reaction catalyzed by L-glutamate dehydrogenase.

a review by Earl Stadtman (26), summarizes the interrelationships in metabolic function involving the enzyme. It is reasonable to assume that some of the chemical and physical effects that I will discuss in this lecture may have some physiological importance in the in vivo function in the mitochondria of the liver (from where the beef liver enzyme which we have studied originates), although very little is known about this aspect at the present time.

A speculative scheme for regulation of activity and state of aggregation, which was suggested by Tomkins and his collaborators in 1965 (29) based on the consequences of the treatment of solutions of glutamic

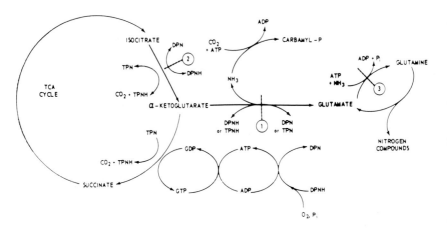

Fig. 3. Interrelation between regulation of tricar-
boxylic cycle and regulation of various bio-
synthetic processes. The dotted lines indicate points of allosteric control (from Reference 26).

dehydrogenase with various low molecular weight substances, is summarized in Figure 4. In Figure 1 we

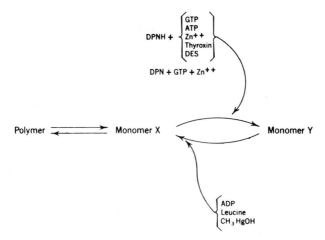

Fig. 4. Dissociation of glutamic dehydrogenase.
DES refers to diethylstilbestiol. See Stadtman (26), after Tomkins *et al.* (29).

saw evidence for a polymerization reaction and a reversible equilibrium between the monomeric enzyme and the polymer. An early observation was that NADH and GTP bring about depolymerization of the enzyme – in other words, shift the equilibrium towards monomer, whereas ADP or leucine abolish and reverse the dissociating effect of these reagents. Furthermore, NADH and GTP bring about inactivation, whereas addition of ADP activates the enzyme. Therefore one of the earliest hypotheses was that one is dealing with two forms of the enzyme, a polymeric form which is active, and a monomeric form which is inactive. This idea had to be modified when more was learned about the state of association of the enzyme at various concentrations. Enzyme assays are usually undertaken at such low protein concentrations, that, even in the presence of reagents which promote association, the enzyme is in the dissociated form; it was indeed very difficult to show whether there was any relationship between the state of aggregation of the enzyme and the enzymatic activity. The original hypothesis was therefore modified by suggesting that one is dealing with two forms of the monomer. One form, "monomer X," was visualized as the active form of the enzyme, and the inactive form was termed "monomer Y." However X apparently has "sticky" surfaces which could attach to each other, while monomer Y was not capable of polymerizing. We therefore have two functions here, one the enzymatic function, and the other one the ability to polymerize. [Much of the discussion in the above paragraph has been reviewed by Stadtman (26), Frieden (9,10) and Tomkins and collaborators (29,30) and I refer to these works for a more detailed exposition of these authors' views.]

To date very little is known about the significance of the ability to polymerize although it may well play a role in higher order in vivo structures of the enzyme. Enzymes from animals on a lower phylogenetic scale, such as GDH from rat liver for instance, show a smaller tendency to associate, although association may be promoted (13) by saturation of the buffered solutions with toluene, a reagent which we

47

showed (22,23) strongly promotes glutamic dehydrogenase association in a mode indistinguishable from the natural concentration-dependent self association. To us, the phenomenon of self association has proved to be an extremely useful tool in allowing us to observe, with methods with which we are familiar and for which we are well equipped, a number of correspondances between macromolecular structure and some well documented enzymatic properties of the enzyme. These are subject to allosteric regulation but no detailed mechanisms have been investigated and the whole subject remains very complicated. It has been possible though, by careful chemical modification of simple amino acids in the active subunit of the enzyme (cf. for instance the work of Colman and Frieden (4,5)) to effect changes sometimes leading to decoupling of the association, regulatory and activity mechanisms. It has also been claimed (29,30) that, in the "inactive" monomer Y form, glutamic dehydrogenase shows enzymatic activity towards alanine and other monocarboxylic acids, albeit at a much lower level. I will not pursue here detailed analysis of these very complex enzymatic processes. Some pertinent reviews have recently been published (9,10,11,27).

Molecular Weight of Oligomer and Subunits

Figure 5 is taken from a paper (8) published with Gordon Tomkins when we first attacked problems related to the molecular weight of the enzyme. From studies by light scattering we obtained the weight average molecular weights at various concentrations. If these are plotted against concentration one can detect a big decrease in molecular weight with decreasing concentration. It was necessary to measure the scattering at very low concentrations in order to extrapolate to the limiting value of the molecular weight. Previously molecular weight determinations by either light scattering or equilibrium sedimentation had been obtained at much higher enzyme concentrations and it had proven very difficult to obtain

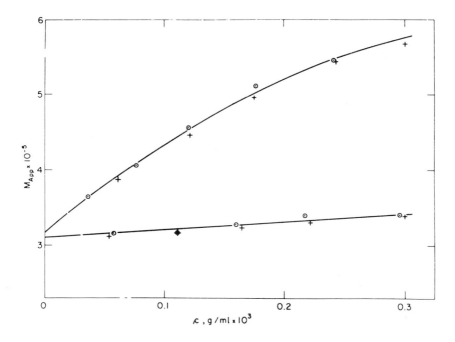

Fig. 5. Apparent weight-average molecular weight of glutamate dehydrogenase as a function of enzyme concentration. Upper curve, dialyzed against 0.2 M phosphate buffer, 10^{-3} M EDTA, pH 7; lower curve, same as above plus 10^{-3} M GTP, 10^{-3} M NADH. Light scattering experiments at 25°C, at 546 nm. Circles and crosses represent independent experiments (Eisenberg and Tomkins (8)).

the exact molecular weight by extrapolation. The lower curve in Figure 5 is also a light scattering curve obtained upon addition of 10^{-3} M GTP and 10^{-3} M NADH to the buffered enzyme solutions. As you remember from Figure 4 these reagents lead to dissociation of the enzyme and at the same time reversibly inhibit the activity. What we now find from this result is a considerable drop in association at finite enzymic concentrations and, upon extrapolation to vanishing concentration, we obtain, within experimental error, the same molecular weight average (313,000) as

in the absence of the regulatory agents. We thus concluded that the basic active unit of the enzyme has a
molecular weight of 313,000 and that the low molecular
weight effectors we studied (GTP plus NADH at 10^{-3} M)
affect the state of association of the enzyme but not
the molecular weight of the monomeric unit. This active unit of 313,000 daltons is also sometimes called
an oligomer, because it is composed of a number of
identical subunits. I will not give you extensive details on the determination of the number of subunits,
since the subject could easily be the topic for a
second lecture. Briefly, if you dissolve the enzyme
in 6 M guanidine hydrochloride solutions (in the presence of some disulfide reducing reagent such as mercaptoethanol, for instance) the noncovalently linked
peptide chains (subunits) fall apart and their molecular weight can be determined by either light scattering or equilibrium sedimentation. The experiments and
their interpretation is a little tricky but one obtains
the correct result if some straightforward procedures
relating to preferential interactions and the problems
raised by the thermodynamics of multicomponent systems
are followed (3,21). Here we were lucky that chemical
evidence had established (1) that the peptide chains
were identical and we were thus determining a well defined molecular weight, and not an average value. We
found a molecular weight of 53,500 for these subunits
(peptide chains) and simple arithmetic now tells us
that the closest integer to the ratio 313,000/53,500 =
5.85 is 6 and that the active monomer is therefore
composed of six noncovalently linked subunits. This
conclusion raised some skeptical brows when submitted
for publication because it seemed to contradict the
current dogma concerning the number of subunits a self-
respecting allosteric enzyme should contain (six was
not a good number; either four or eight would have been
greeted with considerably more applause). Yet the idea
of six subunits in active glutamic dehydrogenase has
withstood the test of time, and has been confirmed by
electron microscopy as we shall see in a moment. The
sequence work of Emil Smith and his collaborators (25),

which has come to fruition after many years of hard
work, sets the molecular weight of the peptide chains
at 56,000, just 4% higher than the result obtained by
the physico-chemical studies in solution.

To summarize, we are dealing with an active oligomer composed of 6 subunits. It has not been possible
to obtain active enzyme units lower than 313,000. When
we speak about reversible dissociation of GDH we always
mean reversible dissociation to the active hexamer; the
six subunits are identical and each carries a full complement of active and regulatory sites. This is not
necessarily so in all other enzyme systems.

A Physical Model for the Active Enzyme

Having established the number of subunits per
enzyme oligomer, interest then turned towards elucida-
tion of the geometry of the arrangement of these subunits. Such information might tell us something about
the mechanism of the polymerization reaction. It is
amazing that at that time, just a few years ago, nothing was known about the geometry of the enzyme and no
proper electron micrographs had been obtained. It is
very strange that one can obtain very beautiful pictures of one enzyme and that the methods simply do not
work for another. We were very pleased, therefore,
when Robin Valentine, in 1968, published (31) some
pictures (Figure 6) taken of this elusive enzyme; he
confirmed earlier observations (12) that the molecule
had a triangular profile: "this suggests two layers
with 3 units in each." Unfortunately further work on
glutamic dehydrogenase by this eminent electronmicro-
scopist was cut short by his untimely death. We knew
by then that the molecule was a hexamer and the hypothesis that GDH had two layers of 3 units, superimposed on each other, fell in quite naturally with our
own result and the physical model we then proposed
(7). The curious thing, when you look at Valentine's
pictures, is that you see only triangles and no indication whatsoever of the layer below it. If the
layers were staggered, as would be the case with the

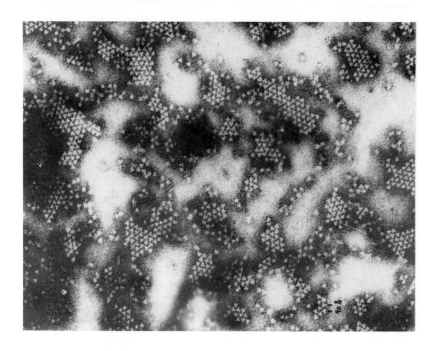

Fig. 6.   Electron micrographs of glutamic dehydrogenase by Valentine (31).

three hydrogen atoms on one carbon atom with respect
to the three hydrogen atoms on the other carbon atom
in an ethylene molecule in a state of minimum energy
(or the three corners of one triangle of a shield of
David, with respect to the three corners of the other
triangle, ✡), then we should catch a glimpse of the
other layer when looking along the three-fold axis of
symmetry of the molecule. Furthermore we should see
different views when viewing along different directions. We built our first model as consisting of two
superimposed trimers. However, on the basis of more
recent evidence from the work of Robert Josephs (14)
undertaken at the Medical Research Council in Cambridge
(Figure 7) we now believe that the two layers are
indeed staggered. Obviously solution studies, or
probing with light having a wavelength of around
5000 Å, cannot distinguish between these two situations.

It may be that the staining procedure used by Valentine, which involved phosphotungstic acid, brought about separation of the active oligomers into halves (trimers), the kind of process we have never been able to duplicate in solution. In some of Joseph's micrographs as

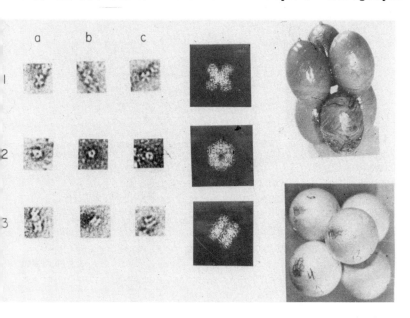

Fig. 7. Columns a to c, electron micrographs of isolated oligomers of glutamic dehydrogenase in various orientations. Next column, computed projections of a model having spherical subunits. These projections are to be compared with micrographs to the left on the same row. At right - two models for glutamic dehydrogenase. The subunits in the lower model are spherical; those in the upper are elliptical (axial ratio 1:1.5). Both models yield similar shadowgraph projections. After Josephs (14).

well there is some indication of a small percentage of triangular molecules. If you look at Joseph's micrographs of intact oligomers (Figure 7) you see that some molecules look like crosses, some like donuts with holes in the middle and some like two layers of material. If you select the model having two layers of spheres, or of ellipsoids, staggered with respect to each other (32 point group symmetry in the form of a triangular anti-prism, in the jargon of the crystallographer) and ask the computer to generate various projections of this model, these correspond exactly to the densities of mass observed. It has not been possible to obtain adequate dimensions from single oligomer molecules with the resolution of which either light scattering or electron microscopy are capable, but we shall see below that it is possible to learn more from the study of the associated forms of the enzyme.

Figure 8 presents further evidence (24) that light scattering and equilibrium sedimentation give corresponding results. The solid points are light scattering results and the squares are individual equilibrium sedimentation runs spanning the error limits associated with these results. Weight average molecular weights, M_w, are given by both experiments. We call these *apparent* M_w's because the so-called second virial coefficient, whose contribution is negligible in this concentration range, have not been taken into consideration. The curves have been calculated on a linear stacking association model in which it is assumed that association proceeds lengthwise along the three-fold symmetry axis and each succeeding oligomer nests snugly (with 60° displacement) in the space offered by the trimeric unit facing it. Actually the only assumption in the calculation is that the infinite association is described by a single association constant or, in other words, that the free energy of adding one oligomer on top of the others is identical and independent of chain length (think of stacking a pile of coins on top of each other, or of the infinite, reversible stacking of nucleotides with a single association constant). Figure 9 represents further evidence for the mechanism proposed. Whereas the molecular weights shown in Figure 8 are weight average molecular

54

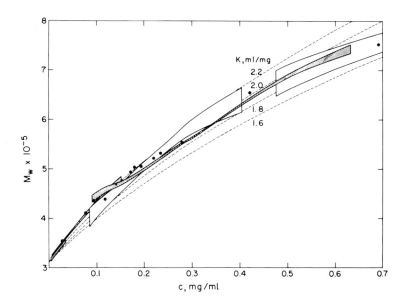

Fig. 8.   Weight average molecular weight (M_w) vs. concentration in 0.2 M Na phosphate buffer, pH 7,
10^{-4} M EDTA, at 20°C; boxes, computed values
from equilibrium sedimentation runs, with estimated error; filled circles, light-scattering
results, also at 20°C; curves, calculated according to reversible infinite linear association (stacking) with single value of association constant K.   From Reference 24.

weights, M_w, it is possible from an analysis of equilibrium sedimentation (though not from light scattering) to derive both a lower - the number average molecular weight, M_n, - and a higher - the z - average
molecular weight, M_z. These averages can be computed
by a rather simple calculation and the result indicates that the same association constant describes all
the measured molecular weight averages. M_n is obtained
quite accurately, but usually the higher average, M_z,
is not give so precisely.
        From the angular dependence of scattering of
light we derive unique information about the size of

55

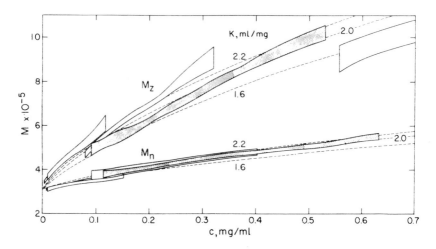

Fig. 9. Z-average (M_z) and number average (M_n) mol-
        ecular weights from equilibrium sedimentation;
        curves and symbols as in Figure 8. From
        Reference 24.

the particles. Usually light scattering, which, as I
have already mentioned, uses a wave length of about
5000 Å, is not very suitable for probing details of
the structure of proteins which have dimensions around
50 Å. You would not find it very helpful to measure
the dimensions of cubes having 50 mm edges with a
yardstick 5000 mm long, with no subdivisions. Of
course if you could stack a number of cubes on top of
each other matters would be improved. But you would
also manage nicely disposing of subdivisions of about
1.5 mm, which would correspond (on this scale) to the
scattering of X-ray with 1.5 Å wavelength.
    In summary, nucleic acids are sometimes too big
for light scattering but proteins are usually too small
to yield indications about the sizes of the particles -
the scattering curve is independent of the scattering
angle. Figure 10 is a plot (8) of the inverse of the
intensity of the scattered light against $\sin^2 (\theta/2)$,
where θ is the scattering angle; at $\theta=0$ the intercept
is proportional to the reciprocal of M_w. The curves

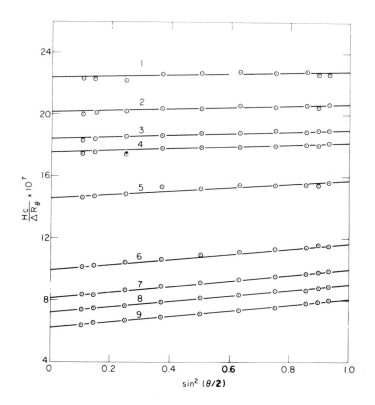

Fig. 10. Angular dependence of reciprocal scattering
function Hc/ΔR(θ) of glutamic dehydrogenase;
0.2 M phosphate buffer, pH 7, 10^{-3} M EDTA,
λ=546 nm, at various concentrations. In
descending order, concentrations (mg/ml) and
R_g (Å) respectively: Curve 1 (0.121,70),
2 (0.180,90), 3 (0.241,108), 4 (0.300,105),
5 (0.572,157), 6 (1.436,234), 7 (2.15, 272),
8 (2.84,281) and 9 (7.86,299). From Reference 8.

in descending order correspond to increasing concentration. At low concentration there is no angular
dependence as befits small protein molecules. With
increasing concentration the intercept decreases, indicating an increasing molecular weight. Also, the

angular dependence becomes more pronounced, indicating increasing asymmetry with increasing association. This observation first led us to the hypothesis that association proceeds in a linear array. If we divide the slopes in Figure 10 by the corresponding intercepts we obtain a quantity proportional to the square of the radius of gyration, R_g, of the particle; for rodlike particles of constant cross-section, R_g^2 is just $12 L^2$ where L is the length of the rod. Figure 11 shows (7) that R_g linearly increases with the molecular weight, therefore providing a consistent picture for the linear association model.

Effect of Toluene on Enzyme Association

On Figure 11 is another piece of information which we used in our work and which was discovered quite accidentally (23), namely that aromatic hydrocarbons such as toluene or benzene bring about a tremendous enhancement of the polymerization reaction, without any loss of biological activity. All the points represented by crosses on this figure have been obtained in aqueous buffer solutions saturated with respect to toluene and it is seen that these points overlap with the other points (obtained in the absence of toluene). Therefore we appear to be dealing with the same structure, both in the absence and in the presence of toluene. We can obtain very high molecular weights in the presence of toluene; this curve extends to 3.5×10^6 daltons and we have obtained molecular weights of active enzyme complexes as high as about 15×10^6 daltons. This is quite unusual for an enzyme system. Yet we do not understand the significance of the toluene effect. We should remember, however, that in the mitochondria of the liver the enzyme is not in the medium which the biochemist usually creates for it in an enzymatic test tube assay. In the *in vivo* situation we certainly have a large number of hydrophilic and, in particular, hydrophobic components which may exert a profound influence on the structure and function of the enzyme (Footnote 7, reference 6).

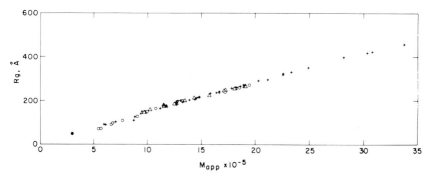

Fig. 11. Radius of gyration R_g from light scattering of glutamic dehydrogenase solutions *vs*. apparent molecular weight; (0) sodium phosphate buffer 0.2 M, pH 7, 10^{-4} M EDTA, 25°C, enzyme concentration range 0.5 to 11 mg/ml; (Δ) same buffer, various temperatures (10-30°C), same range of enzyme concentrations; (+) same buffer, saturated with respect to toluene, enzyme concentration range (0.03 to 0.5 mg/ml); (●) calculated value to the oligomer. From Reference 7.

Another observation is worth mentioning. Whereas ADP, as was indicated earlier, enhances both enzymatic activity and polymerization of the enzyme, the addition of toluene enhances the association tremendously but has no effect on the biological activity. Also, whereas 8-13 moles of toluene are preferentially bound per 53,500 g of associating enzyme, none is bound to the enzyme in the presence of 5 x 10^{-3} M GTP, 5 x 10^{-3} M NADH, when both activity and association are considerably reduced (22).

One final item for the physical chemist; the black point at the left of the plot in Figure 11 is not an experimental point but was calculated to be 46 Å from the model we proposed as corresponding to a single oligomeric unit of molecular weight 313,000. A value of 47 Å has actually recently been reported by Pilz and Sund (19) from small angle X-ray scattering results.

Additional evidence for the linear association
model came from the observation (23) (Figure 12) that
the reduced specific viscosity, η_{sp}/c, $(\eta_{sp} = (\eta-\eta_o)/\eta_o$
where η is the viscosity of the solution and η_o that of
the solvent, and c is the concentration) strongly increases with increasing concentration. From Einstein's
equation, the viscosity contribution due to spherical particles is independent of the size of the particles. Thus, usually protein solutions have very low
viscosities and η_{sp}/c is between 3 and 4. The increase
of η_{sp}/c with enzyme concentration is a strong indication that we encounter increasingly asymmetric particles

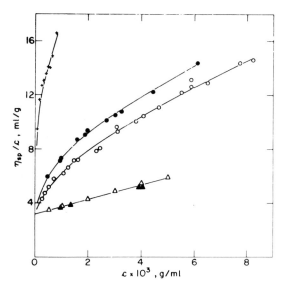

Fig. 12. Viscosity of solutions of glutamic dehydrogenase: (+) in presence of toluene at 20°C;
(●) in 0.2 M phosphate buffer, pH 7, 10^{-4}
EDTA, only, at 20°C; (0) in phosphate buffer
only, at 10°C; (Δ) in phosphate buffer plus
10^{-3} M GTP and 10^{-3} M NADH at 20°C. From
Reference 23.

The lower curve in Figure 12 shows that, as expected,
in the presence of NADH and GTP the increase in η_{sp}/c
is much smaller. The uppermost curve, on the other

hand, dramatizes the effect of toluene on enzyme association: η_{sp}/c increases sharply at extremely low enzyme concentrations.

Quantitative evidence for the linear association may also be obtained from small angle X-ray scattering studies. I have already mentioned that, because of the lower wavelength (1.5 Å as compared to 5000 Å of visible light), this method is sensitive to the dimensions of smaller particles or to the smaller dimensions of larger particles. Thus, in the case of rodlike particles, it is possible to measure a quantity related to the radius of the cross-section of the long rods. Sund *et al.* (28) actually found that the radius of gyration of the cross-section, which is closely related to the radius of the cross-section, is constant at about 30 Å, independent of the state of association of the enzyme particles. By small angle X-ray scattering it is possible also to obtain the mass per unit length of the rods and the result agrees nicely with the result derived from scattering with visible light.

It is of course also possible to visualize the linearly associated particles by electron microscopy (17)(Figure 13). The right field will be discussed in

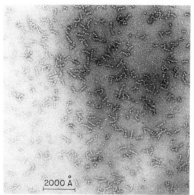 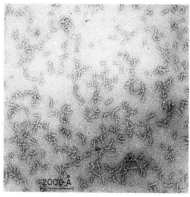

Native   Fixed

Fig. 13.   Electron micrographs comparing native (a) and refractionated crosslinked (b) enzyme. The general appearance of the two fields do not reveal any obvious differences in appearance between fixed and native enzyme. From References 16,17.

the closing part of this talk. It is interesting to note that, along the rod axis, strong (intraoligomer) and weak (interoligomer) bonds (both noncovalent) alternate indicating a definite polarity in the arrangement of the polypeptide chains. Josephs and Borisy (15) have also shown that structures may be obtained on the electron microscope grid which were not observed in solution, in particular tubes formed from four helical intertwined oligomer chains. A great deal of information was derived in this very careful investigation, in particular by using a microscope in which the stage could be tilted from −40° to +40° while looking at the same particle and thereby obtaining a large number of views. Self-assembly of biological macromolecules into specific ordered structures is a topic of major interest in present day studies and takes on a special significance in that it may represent the mechanism of the primary steps in the morphogenesis of whole organisms. Investigation of the control of self-assembly provides an exciting insight into the dynamic nature of the interactions governing many biological processes. Particularly well studied systems have been certain viruses in which the casting off or the reassembly of a multisubunit protein coat around a nucleic acid core is strongly linked to the infectivity and reproduction cycle of the virus.

Crosslinked Linear Enzyme Polymers

The effect of toluene on the enhancement of enzyme association proved to be a very useful tool for creating conditions suitable for intramolecular crosslinking of glutamic dehydrogenase rods by glutaraldehyde, and thereby "fixing" the characteristic rodlike structures, irrespective of changes in environment. Whereas it is usually necessary to perform such crosslinking in concentrated enzyme solution the resulting structures are rather ill-defined, although enzymatic activity may be conserved to some extent. On the other hand in the presence of toluene we could crosslink associated rods of glutamic dehydrogenase

(16,17) at low temperatures and at low enzyme concentrations (about 1 mg/ml) and then, by fractionation on Sepharose columns, obtain crosslinked material of very well-defined molecular weight and molecular structure. The appearance of this material was quite indistinguishable from that of a corresponding preparation of native enzyme (Figure 13).

Conditions chosen for the crosslinking reaction are shown in Figure 14. At the beginning of the reaction the enzyme is fully dissociable (by a dilution

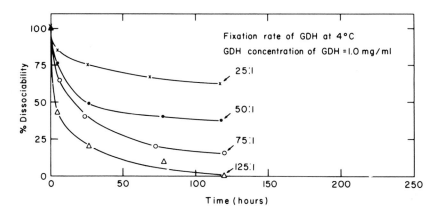

Fig. 14. Plot of dissociability of glutamic dehydrogenase as a function of time with concentration of enzyme at 1 mg/ml and molar excess of glutaraldehyde varied as indicated. From Reference 17.

test) and this property eventually drops to zero. We chose a molar ratio of 75:1 (glutaraldehyde to enzyme oligomer) for the final crosslinking conditions. Chromatography on a Sepharose column (Figure 15) shows that a wide range of weight average molecular weights is obtained, as high as 16×10^6, down to 460,000. For final measurements the material was rechromatographed to eliminate all traces of nonlinear crosslinked material (which can be detected by electron microscopy). Figure 16 is a typical histogram derived by counting particles on electron micrographs, showing

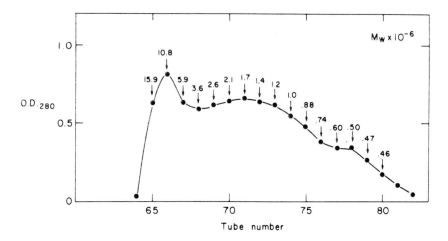

Fig. 15. Chromatogram of fixed glutamic dehydrogenase fractionated on Sepharose 6B. The molecular weights multiplied by 10^{-6} of the fractions are given. From References 16,17.

the distribution of particles in a given sample; M_w derived for this and other samples by light scattering was in rather good agreement. Treatment of these fixed polymers with SDS and electrophoresis on an acrylamide gel column showed that only a few chemical bonds were needed to hold these large structures comfortably together. The very mild crosslinking conditions were used in order to maintain enzymatic activity.

In this talk I will not go into great detail on the relationship between structure and function, the effector sites, the coenzyme sites and the various sites which have been proposed to the enzyme. But let me make a few additional observations of interest. Figure 17 is the Lineweaver-Burk plot of the reciprocal velocity against reciprocal glutamic acid concentration - note the difference in the kinetic behavior between the native and the glutaraldehyde-fixed enzyme. The different points on the curve for the fixed enzyme correspond to different molecular weights ranging from about 0.48×10^6 to 3.6×10^6. The kinetics of the fixed enzyme are independent of chain length, V_{max} of native

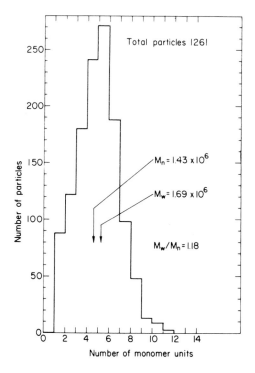

Fig. 16. Typical histograms obtained for rechromatographed fixed enzyme. The histograms show
the number of monomer units per particle of
crosslinked enzyme. From References 16,17.

and mixed enzyme are about the same but the Michaelis
constant K_m for glutamic acid is increased from
1.1×10^{-3} to 1.8×10^{-3} subsequent to fixation. This
indicates a somewhat reduced ability of the polymer to
bind glutamic acid. By contrast (Figure 18), K_m for
NAD (DPN on the figure) remains unchanged upon fixation. Again the behavior of the fixed enzyme is independent of molecular weight. Therefore it appears
that in the linear association process active sites or
coenzyme sites are not blocked in the association
reaction. This seems to be conclusive evidence that
there is no relation between the association *per se*
and the biological activity. In Figure 19 we show

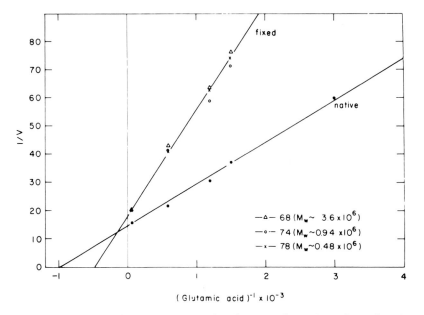

Fig. 17. Double reciprocal plots of initial velocity
with respect to glutamic acid concentration.
The plots compare the activity of three fractions of fixed enzyme (molecular weights are
indicated in the figure) with that of reactive enzyme (enzyme concentration 0.0033 mg/
ml, NAD, 0.43 mM). From References 16,17.

that addition of GTP (in the presence of NADH) inhibits
the fixed enzyme to a significantly smaller extent
than the native enzyme. Again the inhibition is molecular weight independent. Further details and discussion
are given in our forthcoming publication (16).

 This is as far as we have gone so far. From the
point of view of physical chemistry, these molecules are
very interesting; there are strange problems, for example, in the interpretation of the angular dependence of
scattering, but this is not going to add much to the
biology of this problem. On the other hand I feel that
the crosslinking reaction has given us a tool which we
have not exploited to a large enough extent and that
more systematic work can be done and maybe more can be

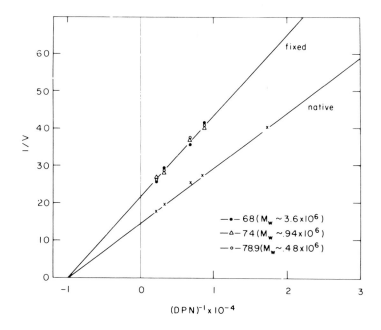

Fig. 18. Double reciprocal plots of initial velocity
          with respect to NAD concentration comparing
          the activity of three fractions of fixed
          enzyme (molecular weights are indicated in
          the figure) with that of native enzyme
          (enzyme concentration 0.0033 mg/ml;
          L-glutamic acid 16.6 mM). From References
          16,17.

found out about the relationship between the active enzyme, the native enzyme and these crosslinked particles.
The whole subject of the interactions of the enzyme is
very large and still controversial (I have not attempted
full literature coverage within the framework of this
talk). I hope that the studies I have discussed here
have contributed some information relating to the behavior of the enzyme, GDH, and may be useful for the
study of other enzyme systems as well. Altogether,
the crosslinking approach is quite fruitful and Burke
and Reisler and coworkers (2,20) are now finding out,
by a similar approach, interesting things about the

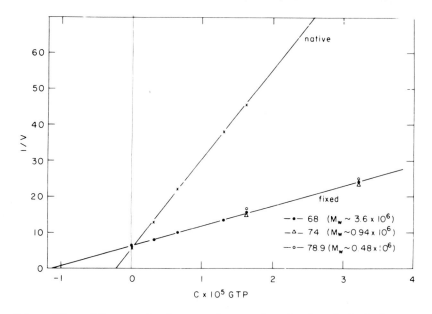

Fig. 19. Plots of the reciprocal of the initial
valocity as a function of GTP concentration.
The plots compare the activity of three
fractions of fixed enzyme (molecular weights
are given in the figure) with that of the
native enzyme (enzyme concentration 0.0016
mg/ml; α-ketoglutarate 8.3 mM; NH_4Cl 20 mM;
NADH 0.1 mM). From References 16,17.

association and mode of action of myosin in Harrington's laboratory at Johns Hopkins.

Acknowledgement

The investigations of the author were supported
in part by Project No. 06-059-1 under the Special
International Research Program of the National Institutes of Health, U.S. Public Health Service.

References

1. Appella, E. and Tomkins, G.M.   The subunits of bovine liver glutamate dehydrogenase: demonstration of a single peptide chain.   J. Mol. Biol. 18, 77 (1966).

2. Burke, M., Reisler, E., Josephs, R. and Harrington, W.F.   Studies on crosslinking of myosin and myosin filaments.   Fed. Proc. 32, 569 (1973).

3. Casassa, E.F. and Eisenberg, H.   Thermodynamic analysis of multicomponent solutions.   Adv. Prot. Chem. 19, 287 (1964).

4. Colman, R.F. and Frieden, C.   On the role of amino groups in the structure and function of glutamate dehydrogenase I. Effect of acetylation on catalytic and regulatory properties.   J. Biol. Chem. 241, 3652 (1966).

5. Colman, R.F. and Frieden, C.   On the role of amino groups in the structure and function of glutamate dehydrogenase II. Effect of acetylation on molecular properties.   J. Biol. Chem. 241, 3661 (1966).

6. Eisenberg, H.   Glutamate dehydrogenase: anatomy of a regulatory enzyme.   Acc. Chem. Res. 4, 379 (1971).

7. Eisenberg, H. and Reisler, E.   A physical model for the structure of glutamate dehydrogenase. Biopolymers 9, 113 (1970).

8. Eisenberg, H. and Tomkins, G.M.   Molecular weight of the subunits, oligomeric and associated forms of bovine liver glutamate dehydrogenase.   J. Mol. Biol. 31, 37 (1968).

9. Frieden, C. Aspects of the effects of purine nucleotides on the control of activity and molecular properties of glutamate dehydrogenases. in The Role of Nucleotides for the Function and Conformation of Enzymes, H.M. Kalckar, H. Klenow, A.M. Petersen, M. Ottensen and J.H. Thaysen (Eds.) Academic Press, New York, p. 194, 1969.

10. Frieden, C. L-glutamate dehydrogenase. in The Enzymes, 2nd ed., vol. 7, P.D. Boyer, H. Lardy and K. Myrbäck (Eds.), Academic Press, New York, p. 3, 1963.

11. Frieden, C. Protein-protein interaction and enzymatic activity. Ann. Rev. Biochem. 40, 653 (1971).

12. Horne, R.W. and Greville, G.D. Observations on the ox-liver L-glutamate dehydrogenase with the electron microscope. J. Mol. Biol. 6, 506 (1963).

13. Ifflander, U., and Sund, H. Association behavior of rat liver dehydrogenase. FEBS Letters 20, 287 (1972).

14. Josephs, R. Electron microscope studies on glutamic dehydrogenase: subunit structure of individual molecules and linear associates. J. Mol. Biol. 55, 147 (1971).

15. Josephs, R. and Borisy, G. Self-assembly of glutamic dehydrogenase into ordered superstructures: multichain tubes formed by association of single molecules. J. Mol. Biol. 65, 127 (1972).

16. Josephs, R., Eisenberg, H. and Reisler, E. Some properties of crosslinked polymers of glutamic dehydrogenase. Biochemistry 12, 4060 (1973).

17. Josephs, R., Eisenberg, H. and Reisler, E. Subunits to superstructures: assembly of glutamate dehydrogenase, in Protein-Protein Interactions,

Proceedings of the 23rd Mosbach Colloquium, B. Jaenicke and E. Helmreich (Eds.), Springer-Verlag, Berlin, p. 57, 1972.

18. Olson, J.A. and Anfinsen, C.B. The crystallization and characterization of L-glutamic acid dehydrogenase. J. Biol. Chem. 197, 67 (1952).

19. Pilz, I. and Sund, H. Studies of glutamate dehydrogenase. The X-ray-small-angle investigation of the beef liver glutamate dehydrogenase oligomer. Eur. J. Biochem. 20, 561 (1971).

20. Reisler, E., Burke, M., Josephs, R. and Harrington, W.F. Studies on crosslinking of myosin and myosin filaments. J. Mechanochem. Cell Motility, in press.

21. Reisler, E. and Eisenberg, H. Interpretation of equilibrium sedimentation measurements of proteins in guanidine hydrochloride solutions. Partial volumes, density increments, and the molecular weight of the subunits of rabbit muscle aldolase. Biochemistry 8, 4572 (1969).

22. Reisler, E. and Eisenberg, H. Solubility of toluene in bovine liver glutamate dehydrogenase solutions and enhancement of enzyme association. Biochim. Biophys. Acta 258, 351 (1972).

23. Reisler, E. and Eisenberg, H. Studies on the viscosity of solutions of bovine liver glutamate dehydrogenase and on related hydrodynamic models; effect of toluene on enzyme association. Biopolymers 9, 877 (1970).

24. Reisler, E., Pouyet, J. and Eisenberg, H. Molecular weights, association, and frictional resistance of bovine liver glutamate dehydrogenase at low concentrations. Equilibrium and velocity sedimentation, light scattering studies, and

settling experiments with macroscopic models of
the enzyme oligomer. Biochemistry 9, 3095 (1970).

25. Smith, E.L., Landon, M., Piszkiewikz, D., Brattin,
W.J., Langley, T.L., Melamed, M.D. Bovine liver
glutamate dehydrogenase: tentative amino acid
sequence; identification of a reactive lysine;
nitration of a specific tyrosine and loss of
allosteric inhibition by guanosine triphosphate.
Proc. Nat. Acad. Sci. U.S. 67, 724 (1970).

26. Stadtman, E.R. Allosteric regulation of enzyme
activity. Advan. Enzymol. 28, 41 (1966).

27. Sund, H. (Ed.) Pyridine Nucleotide Dependent
Dehydrogenases, Springer Verlag, Berlin, 1970.

28. Sund, H., Pilz, I. and Herbst, M. Studies of
glutamate dehydrogenase. The X-ray-small-angle
investigation of beef liver glutamate dehydrogenase. Eur. J. Biochem. 7, 517 (1969).

29. Tomkins, G.M., Yielding, K.L., Curran, J.F.,
Summers, M.R., and Bitensky, M.D. The dependence
of the substrate specificity on the conformation
of crystalline glutamate dehydrogenase. J. Biol.
Chem. 240, 3793 (1965).

30. Tomkins, G.M., Yielding, K.L., Talal, N. and
Curran, J.F. Protein structure and biological
regulation. Cold Spring Harbor Symp. Quant. Biol.
28, 461 (1963).

31. Valentine, R.C. Precongress Abstracts of the
Fourth European Regional Congress of Electron
Microscopy, Bocciarelli (Ed.), vol. 2, Rome, p. 3,
1968.

PROTEINS AS ANTIGENS

David H. Sachs

Immunology Branch
National Cancer Institute
National Institutes of Health
Bethesda, Maryland   20014

It was recognized quite early in the history of
immunology that the injection of naturally occurring
proteins into an animal other than that from which the
protein was derived would lead to the production of
antibodies against that protein. This property, known
as immunogenicity, is so characteristic of proteins
that they were originally thought to be the only substances to which the immune system could react. With
the advent of refined purification procedures, sensitive detection techniques, and the use of adjuvants,
it became clear that all sorts of substances could be
made immunogenic. It was learned that by attaching
low-molecular weight organic molecules, known as haptens, to larger carrier molecules, and immunizing with
the conjugates, that one could raise antibodies to
such haptens, despite the fact that they alone would
not have been immunogenic. Since the antibody-antigen
reactions of proteins are relatively complex compared
to those of haptens, most quantitative investigations
of antibody-antigen reactions have relied heavily on
the use of these haptens.

However, the chemical nature of the antibody-
antigen interaction is now sufficiently well-character-
ized to permit the immunologist, who borrowed so heavily from protein chemistry over the past 20 years, to

be able to reciprocate by providing the protein chemist with useful tools for the study of the structure and function of proteins.

In today's lecture I would like to elaborate on this theme of immunologic approaches to the study of proteins. I shall first review some basic principles of immunochemistry essential to such approaches, then examine some of the specifics of the application of these principles to the antibody-protein interaction. Finally, I would like to discuss in detail some work from our own laboratory on the model protein, staphylococcal nuclease, which I hope will illustrate the essential features of such an immunologic approach.

An antibody may be defined as a serum protein which arises in response to immunization with a given antigen and which reacts with that antigen specifically, *i.e.*, with an association constant greater than, say, 10^5. The reaction of such an antibody with its antigen can be described in the most general case by the equation:

$$S + L \quad\quad SL \tag{1}$$

where S equals the individual combining sites, L equals the ligand (antigen), and SL equals the specific complex. The association constant, $K_{association}$ is then given by:

$$K_{assoc.} = \frac{[SL]}{[S][L]} \tag{2}$$

If we let r equal the number of combining sites to which ligand is bound per antibody molecule at a free ligand concentration c, and if we let n equal the maximum possible combining sites per antibody molecule, then:

$$K_{assoc.} = \frac{r}{(n-r)c} \tag{3}$$

or:

$$r/c = K_{assoc.} n - K_{assoc.} r \tag{4}$$

In order to quantitate the given antibody-antigen reaction, one must be able to measure quantities r and c, and it is at this point that haptens have played such a major role. By the technique of equilibrium dialysis, illustrated in Figure 1, one can readily determine

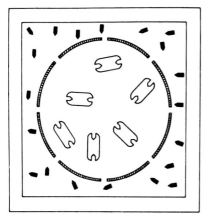 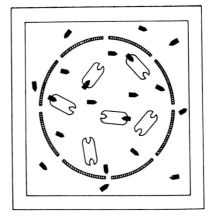

⊋ ⊊ Antibody molecule   ➤ Hapten molecule

Fig. 1.   Schematic diagram of the technique of equilibrium dialysis. The membrane, represented by the broken circle, is permeable to small hapten molecules but not to the larger antibody molecules. The left cell shows the distribution of antibody and hapten initially, and the right cell shows the distribution at equilibrium. From: Pinckard, R.N. and Weir, D.M. in Weir, D.M. Handbook of Experimental Immunology, p. 494, Philadelphia, Pa. (1967) F.A. Davis Co.

values of r for a purified antibody preparation at a variety of values of c for the small, dialyzable hapten. Equation 4 then defines the antibody-antigen interaction in a form suitable for a Scatchard plot (26), of r/c *vs.* r, the negative slope of which defines the association constant for the system, and the x-intercept of which defines n. Such a plot is shown in Figure 2, which compares the antibody-antigen reaction to an enzyme-substrate reaction plotted in a similar fashion. It is apparent that while the enzyme-substrate reaction is defined over the entire binding range by a single, uniform $K_{association}$, the antibody-antigen reaction is characterized by different apparent values of $K_{association}$ depending on the concentration range studied. This has been interpreted as indicating a large degree of heterogeneity in antibody populations due to the presence of antibody molecules of different affinities for the antigen in question.

Such heterogeneity of binding constants is characteristic for antibody populations, except for the myeloma proteins and certain other homogeneous antibody populations, both of which yield linear Scatchard plots (18). Also characteristic for antibodies from hyperimmune animals are x-intercepts approaching r=2, as shown in Figure 2. This indicates a total of two combining sites per antibody molecule, an hypothesis which can now be understood at the molecular level, γG immunoglobulins having been shown to contain two identical binding regions (Fab) per molecule.

Similar heterogeneity of antibody affinities is of course to be expected in the response to protein antigens. However, analysis of the antibody-antigen interaction in this case is further complicated by the multiplicity of antigenic sites, or determinants, on a single protein antigen. While the interaction of each antibody molecule with its determinant must, of course, be described by the same equations we have examined for the case of haptens, the presence of multiple antigenic sites per molecule creates a multicomponent system, making a mathematical description of the interaction very complicated (16). Qualitatively, however, divalent antibodies should be expected to react in only one of three ways with multivalent antigens (Figure 3). Which of the three possible products will be formed in a given reaction depends on the number of antigenic sites per antigen molecule, and on the presence or absence of antibodies specific for each site. If there is only one antigenic site per antigen (or equivalently, if the antibody population is specific for only one

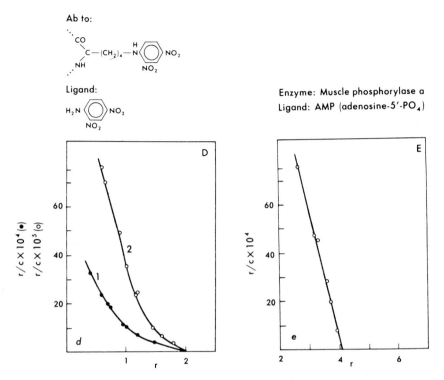

Fig. 2. Comparison of Scatchard plots of the binding of an antibody preparation to its antigen and of the binding of an enzyme to its substrate. From: Eisen, H.N. in Davis, B.D. *et al.* Principles of Microbiology and Immunology, p. 369, New York, N.Y. (1968) Harper and Row.

site) then only the soluble complex antigen-antibody-antigen can be expected. For antigens relatively small compared to γG (molecular weight = 150,000) the complex will not differ greatly in molecular weight from free antibody. If there are two antigenic sites capable of reaction in the system, then one can envision the formation of chains of the form -antibody-antigen-antibody-antigen- which, for reasons of stability, would probably not become very long before either breaking or terminating by ring closure. Such reactions would therefore be expected to yield high

Antigen	Antibodies	Complex in Antibody Excess

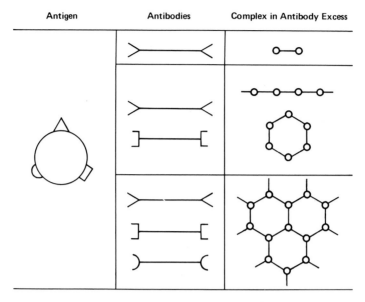

Fig. 3. Schematic diagram of the three possible reactions of antibodies with multivalent antigens. Note that only in the presence of three or more antigenic determinants per antigen molecule and antibodies to each determinant can an effective lattice be formed.

molecular weight aggregates which, however, would likely not precipitate under ordinary conditions. With three or more detectable antigens in the system, one might expect multiple branching to occur, leading to what is known as a lattice (20). Such lattices reach very high molecular weight and uniformly precipitate, which forms the basis for the precipitin reaction, the most versatile tool for studying antibody-antigen reactions of multivalent antigens.

A typical precipitin reaction for a protein (egg albumin) is illustrated in Table 1 and in Figure 4. For a given amount of antibody the precipitate formed by increasing amounts of antigen can be divided into three zones: in the first, the zone of antibody excess, all of the added antigen is incorporated into a lattice, leaving only free excess antibody in the supernatant. In the equivalence zone there are approximately equal

TABLE 1

PRECIPITIN REACTION WITH A PROTEIN AS ANTIGEN

Tube No.	EAc added (mg)	Total protein precipitated (mg)	Antibody precipitated, by difference (mg)	Supernatant test	Ab/Ag in precipitates	
					Weight ratio	Mole ratio
1	0.057	0.975	0.918	Excess Ab	16.1	4.0
2	0.250	3.29	3.04	Excess Ab	12.1	3.0
3	0.312	3.95	3.64	Excess Ab	11.7	2.9
4	0.463	4.96	4.50	No Ab, no EAc	9.7	2.4
5	0.513	5.19	4.68	No Ab, trace EAc	9.1	2.3
6	0.562	5.16	(4.60)	Excess EAc	(8.2)	(2.1)
7	0.775	4.56	(3.79)	Excess EAc	(4.9)	(1.2)
8	1.22	2.58	—	Excess EAc	—	—
9	3.06	0.262	—	Excess EAc	—	—

Each tube contained 1.0 ml of antiserum obtained by injecting rabbits repeatedly with alum-precipitated crystallized chicken ovalbumin (EAc).

Antibody content of precipitates in tubes 6-9 could not be determined by difference, because too much EAc remained in the supernatants. The latter was measured independently in the supernatants of tubes 6 and 7, allowing an estimate to be made of EAc and antibody in the corresponding precipitates (values in parentheses).

Mole ratio Ab/Ag was estimated by assuming molecular weights for EAc and antibody (Ab) of 40,000 and 160,000, respectively.

Based on M. Heidelberger and F. E. Kendall. J. Exp. Med. 62:697 (1935).

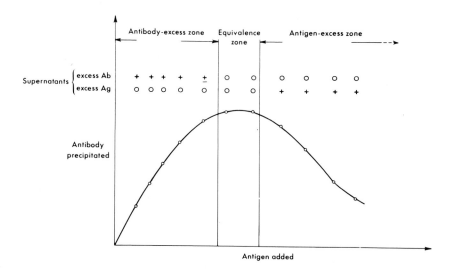

Fig. 4. Classic precipitin curve for an antigen and the corresponding antibodies. From: Eisen, H.N., in Davis, B.D. *et al.* Principles of Microbiology and Immunology, p. 381, New York, N.Y. (1968) Harper and Row.

proportions of antibody and of antigenic determinants, leading to a lattice incorporating all of the antigen and maximal amounts of the antibody. If antigen is further increased, one reaches a zone of antigen excess in which, on the average, insufficient antibodies are present to bind more than one or two antibody molecules per antigen molecule, despite the presence of free antigenic sites on those molecules. The lattice cannot form and the amount of precipitation decreases, free antigen and soluble antibody-antigen complexes being left in the supernatant.

Another simple but elegant method of carrying out this reaction is known as Ouchterlony double diffusion analysis, or precipitation reactions in gel. An example is shown in Figure 5. Here antibody and antigen solutions are placed in wells cut in an agar gel and are allowed to diffuse toward each other in the gel. After diffusion one can observe all three zones of the precipitin curve visually between the wells, a well defined precipitin arc occurring in the appropriate equivalence zone.

While these precipitin analyses can be made quite precise, the data they provide must be regarded as composites of many competing antibody-antigen interactions. In order to use antibodies as quantitative tools, however, one must isolate the interactions of individual antigenic sites. One way of doing this is to isolate a peptide fragment containing the antigenic determinant, and to study the reaction of that fragment with the antibody population, either by measuring a parameter of direct interaction (*e.g.*, fluorescence quenching or complement fixation) or by the inhibitory effect which such antigens may exert upon the equivalence zone precipitin reaction between the entire antibody population and the native antigen. Another approach is to use such peptide fragments to prepare antibodies specific for distinct antigenic sites on the native molecule and then to study the reaction of such specific antibodies with either the fragment or the native protein. Both of these approaches have been used in the immunochemical study of a variety of proteins, including apomyoglobin (4,5,11), lysozyme (3,14), ribonuclease (9,

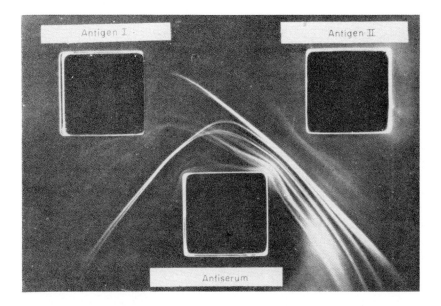

Fig. 5.   Double immunodiffusion in agar, according to
Ouchterlony. The antiserum contains antibodies against many components of the sample
in well marked Antigen II, only a small number of which are also present in the well
marked Antigen I. From: Ouchterlony, O. in
Weir, D.M.: Handbook of Experimental Immunology, p. 702, Philadelphia, Pa. (1967) F.A.
Davis Company.

15), tobacco mosaic virus (7), serum albumins (22,19),
and a variety of synthetic polypeptide antigens (27).
For the purposes of this lecture I shall not attempt
to discuss all of these studies, but will instead concentrate on the details of our own immunochemical
studies of staphylococcal nuclease, which will hopefully illustrate the principles involved. First, however, I would like to discuss briefly the results of
some studies on one of the earliest proteins to be
examined immunochemically, sperm whale apomyoglobin.
I think this will be useful because it illustrates
techniques of general applicability in the study of
proteins as antigens.

Myoglobin is the heme-containing protein of skeletal muscle. Its exceptional abundance in the muscles of aquatic mammals led Kendrew to choose sperm whale apomyoglobin as the subject for the first crystallographic study of a protein's structure (17). A few years later, in deciding on a model protein for an immunochemical study, Crumpton and Wilkinson reasoned that "the definition of the structural features of an antigenic site should be facilitated by a study of the antigenic determinants of a globular protein whose three-dimensional structure and amino acid sequence are known" (11), and they therefore chose sperm whale myoglobin. They produced antisera by immunizing rabbits with purified sperm whale metmyoglobin in complete Freund's adjuvant. They then produced their polypeptide fragments by digesting the protein with chymotrypsin inside of a dialysis sack under positive pressure. Diffusible peptides were thereby separated from indiffusible enzyme and undigested apomyoglobin as they were formed. The peptides were purified by gel filtration and ion exchange chromatography, then tested for ability to interact with anti-sperm whale apomyoglobin antibodies. Since none of these peptides produced precipitates with the antiserum, the authors analyzed antigenic activity by the ability of the peptides to inhibit the precipitation reaction between antibodies and apomyoglobin. Their results are shown in Table 2a, which presents the maximum inhibitions obtained with increasing amounts of the individual peptides. Since the structure and sequence of this antigen were known, Crumpton and Wilkinson could readily determine from which part of the myoglobin molecule their peptide fragments were derived. As can be seen in Figure 6, their results indicate that the isolable antigenic determinants of sperm whale myoglobin (arrows in Figure 6) are on peptides which derive from the corners of the folded polypeptide chain.

Subsequent to Crumpton and Wilkinson's paper, Atassi and coworkers studied a large number of polypeptide fragments of sperm whale apomyoglobin prepared by a variety of techniques of protein cleavage (4,5).

TABLE 2a

*Maximum inhibition of precipitation of
antigen–antibody complex caused by the peptides
isolated from a chymotryptic digest of apomyoglobin*

Maximum inhibition (%) of precipitation of
antigen–antibody complex

	Antiserum WF		Antiserum WH
Peptide	With apomyoglobin	With metmyoglobin	With apomyoglobin
A2	12		
A4	12	8	9
B1	0	8	0
C1a	0		9
C2	0		> 7
D1b	5		
D2	15	7	> 6
D3	0	0	0

TABLE 2b

Molar ratios of inhibitors to antigen

Inhibitor:antigen molar ratio for 50% of
maximum inhibition

	Antiserum WF		Antiserum WH
Inhibitor	With apomyoglobin	With metmyoglobin	With apomyoglobin
A2	12		
A4	50	60	65
B1	—	145	—
C1a	—		150
C2	—		> 290
D1b	70		
D2	80	70	> 300

(From ref. 11)

Their work has confirmed and supplemented that of
Crumpton and Wilkinson. Some of the peptides studied
by Atassi were much larger than the chymotryptic peptides and in fact contained sufficient antigenic determinants to precipitate with antimyoglobin antibodies
(Figure 7). By summing up the antibody activities

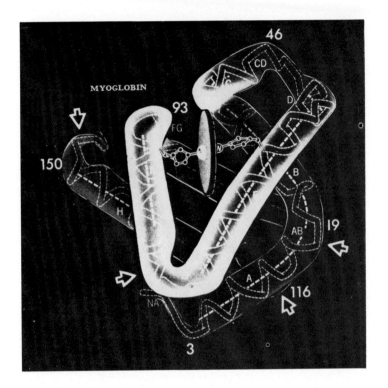

Fig. 6.   Schematic representation of sperm-whale myoglobin. The arrows point to regions implica-
ted as antigenic determinants by the work of
Crumpton and Wilkinson (11).

(either precipitating or inhibitory) of a variety of
these fragments, Atassi and coworkers were able to
account for essentially all of the antibody activity
of their antisera. In no case did they find more antibody binding activity toward the sum of the peptides
than toward the native molecule.

One can conclude from these studies two very general features of native globular proteins as antigens,
both of which are also supported by a variety of other
lines of evidence. First, it is possible for fragments of such proteins to retain antigenic activity
with respect to antibodies prepared against the native

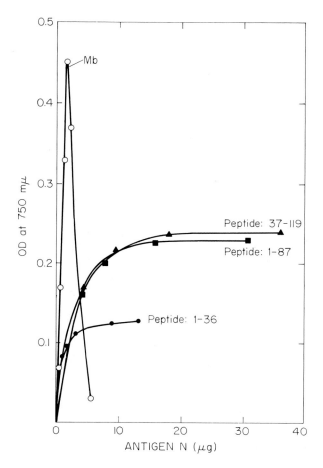

Fig. 7. Quantitative precipitin curves of antimyoglobin antiserum with three myoglobin
peptides. From Atassi, M.Z. and Singhal, R.P.
(5).

protein. Second, essentially all of the antigenic
determinants of such proteins are localized on the
outside of the molecule, indicating that the stimulation of antibody production is generally a feature of
the intact molecule and not of its degradation products.
The data also illustrate two other general features
of the study of proteins as antigens, the basis for which

is less well-understood. The first is the realtively large molar excesses of polypeptide fragments needed to achieve maximum inhibition of precipitation of the native protein. This can be noted in Table 2b, which presents the molar ratio of inhibitor to antigen necessary to achieve 50% maximum inhibition for each peptide. The ratios vary from 12 to >300, indicating a much greater preference of the antibodies for the native antigen than for the polypeptide fragments. From the principles of antibody-antigen binding, which were discussed above, it is not intuitively obvious why this should be so. The second anomalous feature is the shape of the precipitin curve obtained when antisera to native proteins are reacted with polypeptide fragments of those proteins bearing sufficient antigenic determinants to give a precipitation reaction. Generally, such curves do not show a typical zone of decreased precipitation in antigen excess. Examples from the work of Atassi, *et al.*, are shown in Figure 7. It will be noted that the curves obtained with the three peptides shown give a plateau of precipitation even at great molar excess of antigen, where according to the theory of lattice formation, decreased precipitation should be observed. On the basis of our own experiments with staphylococcal nuclease, it is our present feeling that both of these anomalous features may indicate the importance of antigen conformation in the antibody-antigen interactions of proteins. I would therefore now like to discuss these immunochemical studies on nuclease, and to return later to the possible conformational basis of these precipitation anomalies.

Staphylococcal nuclease is an extracellular enzyme produced by *Staphlococcus aureus*. It has been the subject of intensive investigations by Dr. C.B. Anfinsen and numerous collaborators over the past ten years. Both its sequence and its three-dimensional structure are now known (8,10). Its enzymatic activity on DNA is readily measurable by a spectrophotometric assay (12). The enzyme consists of a single chain of 149 amino acids with no disulfide bridges (Figure 8). A feature which makes this enzyme particularly attractive for a

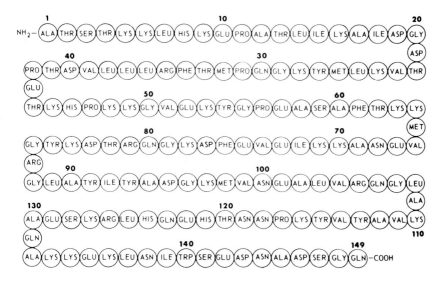

Fig. 8. Amino acid sequence of staphylococcal nuclease, Foggi strain (8).

study of protein conformation is the availability of a variety of complementing systems in which enzymatically inactive fragments of nuclease combine to regenerate activity (1,28-30). There is spectral evidence that this regeneration of activity involves a "folding" process, since the fragments involved are almost totally lacking in the ordered secondary structure of nuclease, while the active complexes regain similar properties to those of the native molecule.

One of the peptide fragments which has been well-studied in this respect is fragment (99 to 149), produced by cyanogen bromide (CNBr) cleavage. In the native enzyme approximately one-half of the sequence represented by this fragment is folded in alpha helix, representing about two-thirds of the total alpha helix of the protein. Since the process of protein folding is thought to involve "nucleation" in regions of ordered secondary structure (2), it was reasoned that the isolation of antibodies specific for this region of nuclease might provide useful reagents for the

study of polypeptide conformation. We therefore set
out to prepare from a hyper-immune goat anti-nuclease
antiserum a population of antibodies specific for a
single antigenic determinant in this helix-rich,
carboxyl-terminal third of nuclease (23). Our approach
to the preparation of these antibodies was to attach
polypeptide fragments of nuclease from this region of
the molecule covalently to Sepharose columns by means
of CNBr activation (13,21). The columns were then
used as solid immunoabsorbents to selectively remove
antibodies capable of reacting with antigenic determinants in these regions. The fractionation is illustrated in Figure 9. Sepharose-nuclease (1-149) was
able to bind all of the antinuclease antibodies in the
serum (about 9 mg/ml). Of these antibodies, 11.5%
(1.0 mg/ml) were specifically bound to the column bearing fragment (99-149).[1] By Ouchterlony analysis the
antibodies obtained at this stage of purification no
longer produced precipitation reactions when incubated
with nuclease, leading us at first to imagine that we
had isolated antibodies to a single antigenic site.
However, in view of the three possible reactions of
multivalent antigens discussed earlier (Figure 3), it
is clear that a population detecting two sites might
also not precipitate. We therefore decided on ultracentrifugation as the criterion of monospecificity of
our antibody population. The binding of one or of
two molecules of nuclease (molecular weight = 17,000)
per antibody molecule (molecular weight = 150,000)
would not be expected to alter the sedimentation behavior of the antibodies greatly, while the binding of

---

[1]According to the nomenclature we have devised
to describe these fractionated antibody preparations,
this population of antibodies is called anti-$(99-149)_n$.
The numbers indicate the relevant sequence of amino
acids of nuclease (Figure 8) and the subscript n indicates that the antibodies were originally prepared
against the native protein.

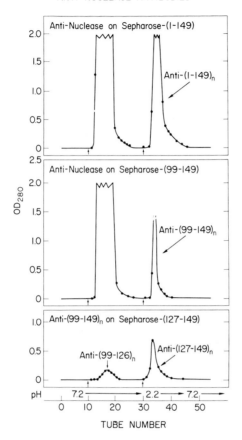

SELECTIVE FRACTIONATION OF
ANTI-NUCLEASE ANTIBODIES

Fig. 9.   Selective fractionation of anti-nuclease
antibodies. For each separation illustrated,
the first peak represents material not capable of being absorbed to the immunoabsorbent
used, and the second peak represents the
specifically absorbed antibodies.

more than one antibody molecule per nuclease molecule
would greatly increase sedimentation velocity. When
anti-$(99-149)_n$ was subjected to an analytical sedimentation velocity ultracentrifugation in the presence of
an equimolar concentration of nuclease, the majority

89

of the protein in the ultracentrifuge cell sedimented much faster than the 7S value expected for γG immuno-globulin.

We therefore refractionated anti-$(99-149)_n$ on another immunoabsorbent column bearing the fragment (127-149) (Figure 9). The first peak from this separation, anti-$(99-126)_n$, contained 26% of the anti(99-149)$_n$ antibodies, or about 3% of the total antinuclease antibodies. When a similar analytical sedimentation velocity ultracentrifugation was performed on this population, only a single species was observed, which sedimented with $S_{20,w}$ of 6.6, indicating the binding of at most one molecule of antibody per molecule of nuclease. At this point, then, we had achieved the preparation of an operationally monospecific antibody population directed toward an antigenic determinant in the carboxyl-terminal third of nuclease. Figure 10 illustrates the position of this antigenic determinant in an artist's representation of the nuclease molecule. We were now ready to study the interaction between these antibodies and the nuclease molecule (24).

Figure 11 shows nuclease assays in the presence and absence of anti$(99-126)_n$. The standard assay involves addition of samples of nuclease to cuvettes containing properly buffered solutions of DNA, with continuous monitoring of OD_{260} on a multiple-sample recording spectrophotometer. In the absence of antibody a linear change of absorption over the first two minutes is observed, corresponding to the expected activity of this amount of nuclease. However, when samples of anti$(99-126)_n$ are added to the cuvettes before addition of nuclease, the slopes of activity are initially similar to that of the control, but fall off well in advance of the time at which they might have been expected to decrease due to limiting substrate. The degree of curtailment of activity increases with increasing initial antibody concentration. In order to explain these results we proposed the following model for the interaction of anti-$(99-126)_n$ with nuclease (Nase) to produce an enzymatically inactive, soluble antibody-antigen complex:

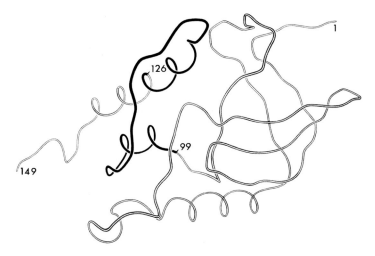

Fig. 10. An artist's representation of the three-
dimensional structure of staphylococcal
nuclease. The drawing was made from a wire
model, based on the X-ray crystallographic
structure. The sequence between amino acids
99 and 126 has been shaded, indicating the
molecular localization of the antigenic determinant of anti-$(99-126)_n$.

$$Ab + Nase \underset{k_{off}}{\overset{k_{on}}{\rightleftharpoons}} AbNase \qquad (5)$$

in which Ab refers to anti-$(99-126)_n$ antibodies, k_{on}
to the second-order association rate constant for the
interaction, and k_{off} to the first-order dissociation
rate constant. From the molar concentration used it
was clear that the experiment had been performed in
large antibody excess, and that the concentrations of
antibody would not have been expected to change greatly
during the reactions. If the proposed model of interaction were correct, then the inactivation process was
expected to follow kinetics which are first order in
nuclease concentration:

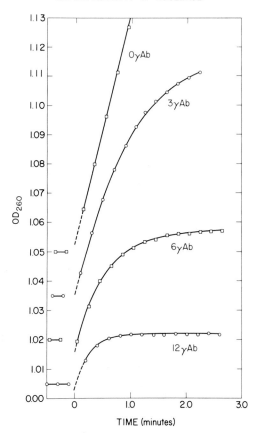

Fig. 11. Inactivation of nuclease by anti-$(99-126)_n$. Three simultaneous activity assays are shown as recorded on a Gilford multiple sample absorbance recorder, with full scale of 0.1 OD unit at 260 nm. At time zero, 0.05 µg of nuclease was added to cuvets containing the indicated amounts of antibody. The points were connected by smooth curves by hand to aid visualization.

$$\frac{d[\text{Nase}]}{dt} = -k_{on} [\text{Ab}][\text{Nase}] \tag{6}$$

or:
$$\frac{d[\text{Nase}]}{[\text{Nase}]} = -k_{on} [\text{Ab}] \, dt \tag{7}$$

which integrates to:

$$\log[\text{Nase}] = \log[\text{Nase}]_o - k_{on}[\text{Ab}]t \tag{8}$$

To test this hypothesis the activity curves were transposed to a semilogarithmic plot. Figure 12 shows data plotted in this fashion. The fact that these data could be fitted to straight lines which, on extrapolation to zero time, originated at the same activity as a control sample of nuclease to which no antibody had been added, supports the validity of this model for the interaction. As can be seen from equation 8, the slope of the lines in Figure 12 should describe the quantity k_{on} (Ab). The values obtained at a variety of antibody concentrations were in close agreement and provided a value of $4.1\pm0.1\times10^5$ M^{-1} sec^{-1} for k_{on}.

In addition to providing the association rate constant for the interaction of anti-$(99-126)_n$ with nuclease, this analysis also provides a rapid and extremely sensitive assay for free antibody: an aliquot of nuclease is added to a standard assay mixture containing an unknown concentration of antibody, and then the rate of decrease in nuclease activity is measured. Since the slope of a plot of the logarithm of nuclease activity *vs.* time measures the quantity k_{on} (Ab), one can readily calculate the unknown antibody concentration. The entire assay takes about two minutes and determines antibody concentrations as low as 10^{-8} M. We shall refer further to this assay later on.

By permitting anti-$(99-126)_n$ antibodies to reach equilibrium with a variety of concentrations of nuclease and then assaying the remaining free nuclease, we were also able to determine the parameters necessary for a Scatchard plot of the interaction according to equation

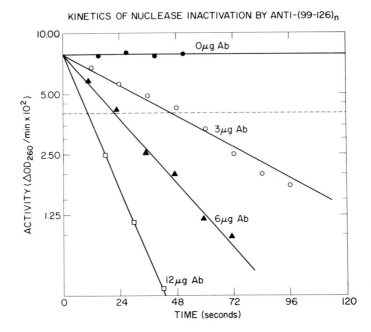

KINETICS OF NUCLEASE INACTIVATION BY ANTI-(99-126)$_n$

Fig. 12. Kinetics of nuclease inactivation by anti-(99-126)$_n$. A semilogarithmic plot of activity *vs*. time for assays of 0.05 µg of nuclease in the presence of the indicated amounts of antibody. The dotted line represents one-half of the initial activity.

4. The Scatchard plot obtained is shown in Figure 13. There are several possible explanations for the shape of this plot, the details of which will not be discussed here (24). Suffice it to say that under the conditions of antibody excess used in all of our kinetic studies, equilibrium would occur at values of r less than unity. The relevant $K_{association}$ under such conditions can be determined adequately from the steep portion of the plot as 8.3×10^8 M^{-1}. The value of k_{off} is then given by:

$$k_{off} = \frac{k_{on}}{k_{association}} \qquad (9)$$

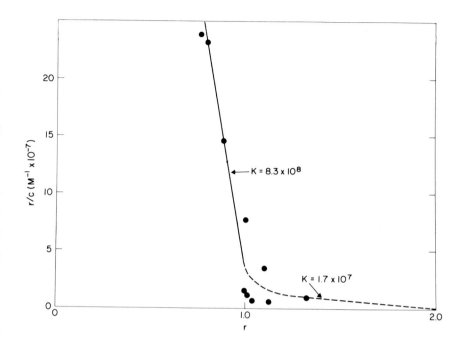

Fig. 13. Scatchard plot. The binding of nuclease by anti-(99-126)$_n$ is plotted as r/s vs. r, values of r being calculated as measured concentrations of bound nuclease per total antibody concentration in equilibrium mixtures. The dotted portion of the curve has been drawn so as to intersect the abscissa at r=2.

or 4.9×10^{-4} M^{-1}, completing the quantitation of the reaction constants of equation 5.

We were now in a position to use this antibody population in the study of conformational equilibria of fragments of nuclease (25). Two lines of evidence led us to believe that the native conformation of the antigenic determinant detected by anti-(99-126)$_n$ might be distinguished from other possible conformations by our

antibody population: First, antibodies prepared by
immunizing goats with the disordered fragments (1–126)
and (99–149), labeled anti-(1–126)$_r$ and anti-(99–149)$_r$,
respectively, showed very different precipitation patterns with nuclease than did antinuclease (Figure 14).
Furthermore, the presence of ligands known to stabilize
the native conformation of nuclease had little effect
on the total precipitable antibody at equivalence for
anti-(1–149)$_n$, but produced marked inhibition of precipitation for anti-(1–126)$_r$ and anti-(99–149)$_r$. This
indicated that many of the antigenic determinants
recognized by the antibodies against the fragments are
present only in the "unfolded" or "non-native" conformation of nuclease. Similarily, the absence of inhibition of precipitable anti-(1–149)$_n$ by ligands suggested
that very little, if any, of these antibodies were
directed towards determinants other than those present
in the native conformation of nuclease.

Second, inactivation studies showed almost no
overlap in the specificity of anti-(99–149)$_n$ and anti-
(99–149)$_r$. Anti-(99–149)$_r$ produced no curtailment of
nuclease activity under conditions in which anti-(99–
149)$_n$ led to complete inactivation of nuclease in less
than one minute. Representative curves of these in-
activations are shown in Figure 15.

We therefore concluded that anti-(99–126)$_n$ reacts
specifically with the *native* conformation of the determinant of (99–126). The ability of these antibodies to bind to disordered fragments of nuclease, such
as (50–149) and (99–149), could then be interpreted by
a simple model involving two simultaneous equilibria.
Polypeptide fragments of nuclease were presumed to
exist in solution in a conformational equilibrium between a variety of disordered or random conformations
(P_r) and the native conformation (P_n). The proposed
equilibrium is illustrated schematically for the fragment (99–149) in Figure 16. The equilibrium can be
described formally as:

$$P_r \rightleftarrows P_n; \quad K_{conf.} = \frac{[P_n]}{[P_r]} \tag{10}$$

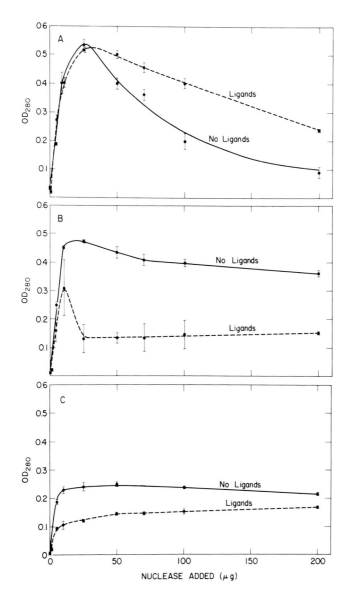

Fig. 14. Quantitative precipitation reactions in the
absence (——) and presence (- - -) of the
ligands pdTp and Ca^{++}.   (A) anti-$(1-149)_n$;
(B) anti-$(1-126)_r$; (C) anti-$(99-149)_r$.   Each
antibody preparation was reacted with increasing amounts of nuclease.

97

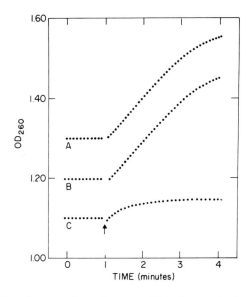

Fig. 15. Conformational specificity of inactivating antibodies. Three simultaneous activity assays are shown as recorded on a Gilford multiple sample absorbance recorder. The cuvette corresponding to the uppermost curve (A) received no antibody, the second cuvette (B) received 18 µg of anti-$(99-149)_r$, and the third cuvette (C) received 6 µg of anti-$(99-149)_n$. At the time indicated by the arrow, 0.05 µg of nuclease was added to each cuvette.

in which it is assumed that $[P_r]$ refers to the sum of the concentrations of all disordered conformations, and K_{conf} is an overall constant defined by this conformational equilibrium at the antigenic site. Antibodies to the native protein are presumed to react effectively only with the form P_n, according to the equation:

$$Ab + P_n \rightleftarrows AbP_n; \quad K_{association} = \frac{[AbP_n]}{[Ab][P_n]} \quad (11)$$

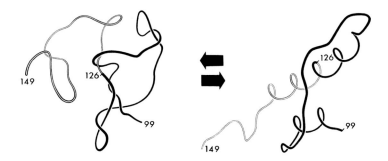

Fig. 16. Artist's representation of the postulated spontaneous, reversible folding of the nuclease fragment (99–149) in solution. The "native format," represented on the right, corresponds to the conformation of this portion of the molecule in intact, native nuclease, based on the X-ray crystallographic structure (10).

Since the antigenic determinant of P_n is, by definition, identical to the corresponding antigenic determinant of nuclease, the association constant for this interaction is assumed to be equal to the experimentally determined association constant for the reaction of these antibodies with nuclease. We then have:

$$\frac{[AbP_n]}{[P_r]} = K_{conf} \, K_{association} \, [Ab] \qquad (12)$$

or

$$K_{conf} = \frac{[AbP_n]}{K_{association}[P_r][Ab]} \qquad (13)$$

For those conformational equilibria for which equation 10 lies far to the left (*i.e.*, low K_{conf}), $[P_r]$ is an adequate approximation to the total polypeptide concentration, $[P_t]$.[2]

To determine the value of the proposed conformational equilibrium constant, K_{conf}, for a polypeptide fragment of nuclease, increasing concentrations of the peptide (P_r) were incubated with anti-$(99-126)_n$, followed by addition of a known quantity of nuclease. From the logarithmic fall-off of nuclease activity the quantity [Ab] was determined, and the quantity (AbPn) was calculated by subtraction from the total anti-$(99-126)_n$. Values of K_{conf} were then calculated from equation (14) and those obtained for the two polypeptide fragments (50-149) and (99-149) are shown in Table 3. It is not surprising from the magnitude of the K_{conf} for both of these fragments that the "folded" fractions of these peptides would be too small (about 0.02%) to be detected by spectral measurements.

Further evidence that K_{conf} does indeed reflect the degree of "nativeness" of the polypeptide in solution was obtained by the use of a synthetic fragment (6-43), an incomplete synthetic analogue of the native fragment (6-48). As was mentioned earlier, native (6-48) is capable of binding to (50-149) in solution to regenerate native structure and activity of nuclease. The synthetic analogue had been shown to bind to (50-

---

$$2 \quad K_{conf} = \frac{[AbP_n]}{K_{association}[Ab][P_t-(AbP_n)]-[AbP_n]} \quad (14)$$

is a more general equation for K_{conf}, which can readily be derived from equations 12 and 13 and the relationship

$$[P_t] = [P_r]+[P_n]+[AbP_n] \quad (15)$$

For low K_{conf} equation 14 can be reduced to equation 13.

149) and to induce tertiary folding, but not to re-generate activity.[3] The values obtained for K_{conf} of fragment (50-149) in the presence of increasing concentrations of synthetic (6-43) are also shown in Table 3. It is apparent from these values that the "folding" of 50-149 induced by the synthetic fragment was reflected in the measurements of K_{conf}. It would thus appear that this immunologic approach may provide meaningful parameters of protein conformation in situations requiring great sensitivity.

TABLE 3

K_{conf} *of nuclease fragments*

Fragment(s)	Concentration of fragment(s) (μM)	K_{conf} ($\times 10^4$)	% free P_t as P_n
(99-149)	0.6	2.20	0.022
(99-149)	2.0	2.02	0.020
(99-149)	2.6	2.29	0.023
(99-149)	7.8	1.47	0.015
(99-149)	6.5	1.51	0.015
		Avg. 2.0	
(50-149)	2.4	2.0	0.020
(50-149)	4.7	2.6	0.026
		Avg. 2.3	
(50-149) + Syn (6-43)	2.4 0.5	6.5	0.065
(5μ-149) + Syn (6-43)	2.4 1.0	18	0.180
(50-149) + Syn (6-43)	2.4 1.9	57	0.560

I would like to return now to the anomalies of inhibition and precipitation of polypeptide fragments noted earlier, and to illustrate how consideration of antigenic conformation similar to those which we have developed for the study of nuclease may help to explain

---

[3]This absence of activity was necessary because our assay for free antibody requires the inhibiting antigen to not have enzymatic activity of its own.

these anomalies. An explanation of the relatively high molar excess of fragment necessary to produce precipitation inhibition may already be obvious, since it was implicit in our treatment of the binding of anti-$(99-126)_n$ to nuclease fragments. Recall from equation 12 that the combination of an anti-"native" antibodies with a fragment undergoing a conformational equilibrium will be given by:

$$K_{association} \; K_{conf} = \frac{[AbP_n]}{[Ab][P_r]} \tag{16}$$

The association constant for the entry of these same antibodies into the precipitating lattice (P_{n-lat}) would be given by:

$$K_{association} = \frac{[AbP_{n-lat}]}{[Ab][P_{n-lat}]} \tag{17}$$

Thus, dividing equation 16 by equation 17:

$$\frac{[AbP_n]}{[AbP_{n-lat}]} = \frac{[P_r]K_{conf}}{[P_{n-lat}]} \tag{18}$$

For 50% inhibition of precipitation we can let $[AbP_n] = [Ab-P_{n-lat}]$, giving:

$$\frac{[P_r]}{[P_{n-lat}]} = \frac{1}{K_{conf}}$$

Thus, a molar excess equal to $\frac{1}{K_{conf}}$ (e.g., 5000-fold for fragment (99-149)) will be required.

This analysis, by the way, should provide a relatively simple method for estimating K_{conf} of a polypeptide. However, the value obtained would generally be a rough estimate, since the analysis can only be applied rigorously to a fragment containing a single antigenic determinant, and since the term $[P_{n-lat}]$ cannot be precisely quantitated.

A similar explanation can be invoked to explain the second anomaly, the failure of excess antigen to inhibit precipitation, in the case of larger polypeptide fragments of protein antigens. The zone of antigen excess requires competition between free antigenic determinants in solution and free determinants bound in the precipitating lattice. Considering the conformational transitions of the polypeptide as cooperative phenomena, one might suppose that once an antibody molecule has bound to one native antigenic determinant of the fragment, other determinants on the same fragment might likewise be "frozen" in their native formats.[4] In other words, the native or random formats are properties of the fragment as a whole rather than just of the region composing the antigenic determinant. The "frozen" antigenic determinants would then be expected to compete much more favorably for further antibody than would antigenic determinants on free polypeptide, effectively eliminating the competition necessary for formation of a zone of antigen excess. One might even consider the extent of a plateau phenomenon as a measure of the degree of cooperativity in the folding of the fragment in question. The long, flat plateaus seen with apomyoglobin (Figure 7) would then indicate a high degree of cooperativity, as might be expected for peptides which are known to be capable of forming alpha helix (6).

This concludes the studies and interpretations which I wished to present in today's lecture on proteins as antigens. In summary, we have reviewed some basic quantitative principles of immunochemistry and have seen that it is possible to apply these same principles to the study of protein antigens. The examples described illustrate the methods available for isolating individual antibody-antigen reactions in

---

[4]We have used the term "native format" to designate an antigenic determinant of a native protein, all the components of which can be generated by a limited length of the polypeptide chain.

protein systems, via the isolation of either the antigenic determinant or the antibodies to a distinct antigenic determinant, and for studying the reactions quantitatively. We have seen that the conformational properties of proteins must be considered as an essential part of their antigenicity, and that such consideration may explain some of the otherwise perplexing anomalies in antibody-protein interactions. Finally, we have seen that when one does include conformation as an explicit variable in the mathematical description of the antibody-antigen interaction, one can then use that interaction as a sensitive means of quantitating the conformational equilibria of polypeptide and protein antigens.

References

1. Andria, G., Taniuchi, H. and Cone, J.L. The specific binding of three fragments of staphylococcal nuclease. J. Biol. Chem. 246, 7421 (1971).

2. Anfinsen, C.B. Synthetic analogues of staphylococcal nuclease: studies of activity and conformation. Proceedings of the 23rd Internatinal Congress on Pure and Applied Chemistry 7, 263 (1971).

3. Arnon, R. A selective fractionation of anti-lysozyme antibodies of different determinant specificities. Eur. J. Biochem. 5, 583 (1968).

4. Atassi, M.Z. and Saplin, B.J. Immunochemistry of sperm whale myoglobin I. Biochemistry 7, 688 (1968).

5. Atassi, M.Z. and Singhal, R.P. Immunochemistry of sperm whale myoglobin VIII. Biochemistry 9, 3854 (1970).

6. Atassi, M.Z. and Singhal, R.P. Conformational studies on modified proteins and peptides. J. Biol. Chem. <u>245</u>, 4122 (1970).

7. Benjanimi, E., Young, J.D., Shimizu, M. and Leung, C.Y. Immunochemical studies on the tobacco mosaic virus protein I. Biochemistry <u>3</u>, 1115 (1964).

8. Bohnert, J.L. and Taniuchi, H. The examination of the presence of amide groups in glutamic acid and aspartic acid residues of staphylococcal nuclease. J. Biol. Chem. <u>247</u>, 4557 (1972).

9. Brown, R.K. Studies on the antigenic structure of ribonuclease. J. Biol. Chem. <u>237</u>, 1162 (1962).

10. Cotton, F.A. and Hazen, E.E. Staphylococcal nuclease X-ray structure, in <u>The Enzymes</u>, Vol. IV, P.D. Boyer (Ed.), Academic Press, p. 153, 1971.

11. Crumpton, M.J. and Wilkinson, J.M. The immunological activity of some of the chymotryptic peptides of sperm-whale myoglobin. Biochem. J. <u>94</u>, 545 (1965).

12. Cuatrecasas, P., Fuchs, S. and Anfinsen, C.B. Catalytic properties and specificity of the extracellular nuclease of *Staphylococcus aureus*. J. Biol. Chem. <u>242</u>, 1541 (1967).

13. Cuatrecasas, P. Insulin-Sepharose: immunoreactivity and use in the purification of antibody. Biochem. Biophys. Res. Commun. <u>35</u>, 531 (1969).

14. Fujio, H., Sakato, N. and Amano, T. The immunological properties of region specific antibodies directed to hen egg-white lysozyme. Biken J. <u>14</u>, 395 (1971).

15. Isagholian, L.B. and Brown, R.K. Interaction of a peptide with antibody to oxidized ribonuclease. Immunochemistry 7, 167 (1970).

16. Kabat, E.A. Kabat and Mayer's Experimental Immunochemistry, 2nd edition, Thomas, Springfield, Ill.

17. Kendrew, J.C. and Pavich, R.G. The crystal structure of myoglobin III. Sperm whale myoglobin. Proc. Roy. Soc. A. 238, 305 (1956).

18. Krause, R.M. The search for antibodies with molecular uniformity. Advances in Immunology 12, 1 (1970).

19. Lapresle, C. and Webb, T. Isolation and study of a fragment of human serum albumin containing one of the antigenic sites of the whole molecule. Biochem. J. 95, 245 (1965).

20. Marrack, J.R., Hoch, H. and Johns, R.G.S. The valency of antibodies. Br. J. Exp. Pathol. 32, 212 (1951).

21. Omenn, G.S., Ontjes, D.A. and Anfinsen, C.B. Fractionation of antibodies against staphylococcal nuclease on Sepharose immunoadsorbents. Nature 225, 189 (1970).

22. Porter, R.R. The isolation and properties of a fragment of bovine serum albumin which retains the ability to combine with rabbit antiserum. Biochem. J. 66, 677 (1957).

23. Sachs, D.H., Schechter, A.N., Eastlake, A. and Anfinsen, C.B. Antibodies to a distinct antigenic determinant of staphlococcal nuclease. J. Immunol. 109, 1300 (1972).

24. Sachs, D.H., Schechter, A.N., Eastlake, A. and Anfinsen, C.B. Inactivation of staphylococcal

nuclease by the binding of antibodies to a distinct antigenic determinant. Biochemistry 11, 4268 (1972).

25. Sachs, D.H., Schechter, A.N., Eastlake, A. and Anfinsen, C.B. An immunological approach to the conformational equilibria of polypeptides. Proc. Nat. Acad. Sci. U.S.A. 69, 3790 (1972).

26. Scatchard G. The attraction of proteins for small molecules and ions. Ann. N.Y. Acad. Sci. 51, 660 (1949).

27. Schechter, B., Schechter, I. and Sela, M. Antibody combining sites to a series of peptide determinants of increasing size and defined structure. J. Biol. Chem. 245, 1438 (1970).

28. Taniuchi, H., Anfinsen, C.B. and Sodja, A. Nuclease-T: an active derivative of staphylococcal nuclease composed of two noncovalently bonded peptide fragments. Proc. Nat. Acad. Sci. USA 58, 1235 (1967).

29. Taniuchi, H. and Anfinsen, C.B. An experimental approach to the study of the folding of staphylococcal nuclease. J. Biol. Chem. 244, 3864 (1969).

30. Taniuchi, H. and Anfinsen, C.B. Simultaneous formation of two alternative enzymically active structures by complementation of two overlapping fragments of staphylococcal nuclease. J. Biol. Chem. 246, 2291 (1971).

THE NIH SHIFT AND ITS IMPLICATIONS FOR THE MECHANISM OF BIOLOGICAL OXIDATIONS

Bernhard Witkop

Laboratory of Chemistry
National Institute of Arthritis, Metabolism,
and Digestive Diseases
National Institutes of Health
Bethesda, Maryland   20014

The need for a rapid assay of phenylalanine hydroxylase (23), a soluble enzyme system that converts phenylalanine to tyrosine, which is missing in the either *classical* or *typical* hereditary genetic defect phenylketonuria (18), led to the discovery of an unexpected phenomenon, the *NIH-shift* (21). As expected, hydroxylating enzymes liberate stoichiometric amounts of tritium from tritiated substrates (9), such as *trans*-4-^{3}H-L-proline (27), 3,5-^{3}H-L-tyrosine (43), β, β-^{3}H-dopamine or tyramine (8) and permit a rapid assay of, *e.g.*, proline, tyrosine and dopamine hydroxylases, respectively. In this case, however, tyrosine was formed from 4-^{3}H-phenylalanine, 1, with 95% of tritium retained 2 (23).

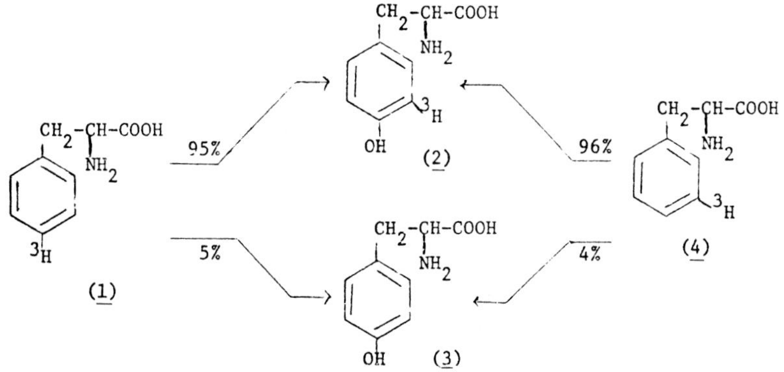

109

3-^{3}H–Phenylalanine, $\underline{4}$, is enzymatically hydroxyl-
ated to tyrosine, $\underline{3}$, and 3-^{3}H–tyrosine, $\underline{2}$, in the same
ratio, 1:19, indicative of a common intermediate, $\underline{5}$,
which aromatizes with loss of a proton to give $\underline{7}$,

rather than loss of a ^{3}H$^+$ to yield $\underline{6}$. If tritium is
flanked by deuterium, $\underline{8}$, the tritium isotope effect,

operative in the aromatization of $\underline{9}$, decreases to a
ratio of 6:4 in the formation of $\underline{10}$ and $\underline{11}$ (R = OCH$_3$)
in which only 20% of tritium is lost and the ratio of
7:6 is 80% *versus* 20% (49). In 3,5-^{2}H-4-^{3}H–acetanilide
($\underline{8}$, R = NHAc) the retentions *in vitro* and *in vivo* differ,
but the ratio of ^{3}H retention in the H(I) and ^{2}H com-
pound (II) is the same (Table 1).

TABLE 1

	Retention of ^{3}H in 4-^{3}H–acetanilide (%) I	Retention of ^{3}H in 3,5-^{2}H-4-^{3}H acetanilide (%) II	Ratio ^{3}H retention I/II
in vivo	40	26	1.54
in vitro	62	40	1.56

When the o– and o–'–positions of the resultant
phenolic product are nonequivalent, the NIH-shift be-

comes selective: 5-hydroxy-4-^{3}H-L-tryptophan, 13, is formed by tryptophan hydroxylase both from 5-^{3}H-, 12, as well as 4-^{3}H-L-tryptophan, 14, with identical retention of tritium (50).

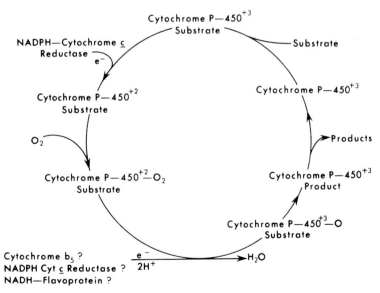

Liver microsomes contain a highly integrated membrane-bound diversified hydroxylating system, the cytochrome P-450 system, represented in the simplified Fig. 1, which lacks the comparative specificity of the soluble phenylalanine and tryptophan hydroxylases.

Fig. 1. Mechanism of Cytochrome P-450 Enzyme Systems.

These hydroxylases are *monooxygenases*, *i.e.*, $^{18}O_2$ is utilized in such a way that one equivalent enters the product, the other the medium. In *dioxygenases*, such as the 3 or 4 different tryptophan-2,3-dioxygenases studied by Hayaishi (24), both oxygen atoms enter the product. Until a few years ago the two types of enzymes seemed to have little in common. Recent evidence points to similar intermediates, ternary complexes and acceptance of superoxide anion as a substitute for active oxygen (24).

Xenobiotic substrates, such as nutrients, steroids and practically all drugs are, as a rule, "detoxified" in the liver by introduction of hydroxyl groups and further reactions to increase water solubility and promote renal filtration and urinary excretion. The hydroxylations of aromatic substrates follow the rules of electrophilic aromatic substitution and should accordingly proceed by addition of active oxygen to the π-electron system: toluene, for instance, besides much benzyl alcohol, is converted to o-, m- and p-cresols in a ratio of 59:13:28 (53). o-Hydroxylation of 2-^{2}-H-anisole, 15, yields o-cresol, 16, with 60% migration

(15)   (16)   (17)   (18)

and retention of deuterium, while m-hydroxylation of 3-^{2}H-chlorobenzene, 17, furnishes deuterated m-chlorophenol with 24% migration and retention (12). o-Hydroxylation is also a well-known phenomenon in plants: 2,6-^{3}H-benzoic, 19, acid retains 15% tritium in the conversion to salicylic acid, 20, whereas o-hydroxylation

(19)   (20)

of the comparable cinnamic acid in the same plant proceeds with 92% retention (17).

The influence of the substituent, R, on the extent
of migration and retention is particularly strong when
R is capable of stabilizing an intermediate cationoid
cyclohexadiene, 21, by contribution of a lone pair of
electrons, 22. The cross-conjugated cyclohexadienol,
22, in contrast to the sequence 24 → 25, can aromatize
only with loss of the hydrogen, *H, originally present
in the p-position. This is the reason why the metab-

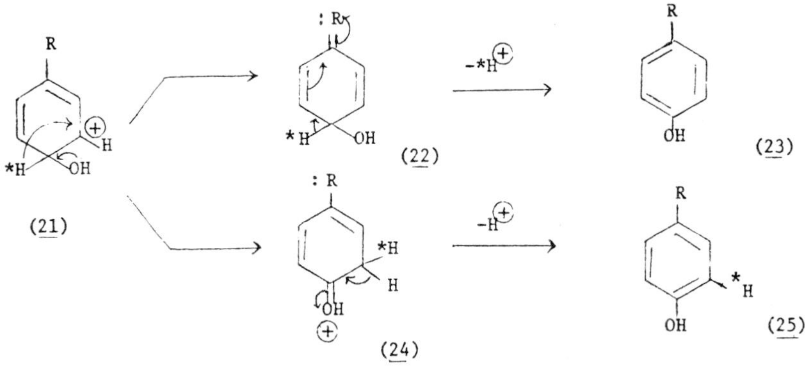

(22)   (23)   (21)   (24)   (25)

olism of anilides shows low retentions of deuterium
(30-40%) and tritium (40-60%) *in vivo* and *in vitro*
(11,13,14).

An experimental approach to the cationoid intermediate, 21, is possible. After administration of
4-^{2}H-chlorobenzene, rabbits excrete the optically active (-)-*trans*-6-^{2}H-5,6-dihydroxy-3-chloro-1,3-cyclo-
hexadiene, 26. The acid-catalyzed dehydration of this

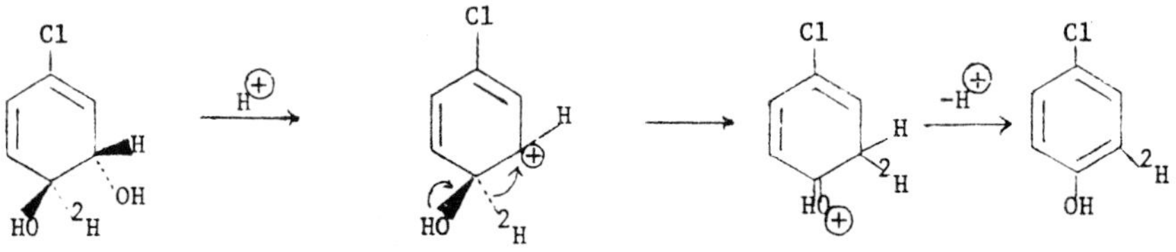

(26)   (27)   (28)   (29)

glycol goes through the cationoid intermediate, 27,
and the cyclohexadienone, 28, which rearranges to

113

2-^{2}H-4-chlorophenol, 29, with 25% retention of deuterium (30). Although the glycol, 26, shows the NIH-shift, it cannot be the primary product of enzymatic oxidation.

Glycols are known products of the enzymatic oxidation of steroids and fatty acids. They result from stereospecific opening of precursor epoxides. When, for instance, cis- or trans-9,10-oxido-steric acids were hydrated with $H_2^{18}O$ and an enzyme from pseudomonas, the isotopic label in the resulting 9,10-dihydroxystearic acids was invariably in C-10 (46).

The comparable "arene oxide," 30, in valence

(31)　　　(30)　　　liver microsomes　　　[α]D-250° (32)　　　(33)
GSH
liver supernate

tautomerism with the oxepin, 31, was found to be a substrate for a solubilizable "epoxide hydrase," (32,47) present in microsomal fractions, and to form the optically active (34) trans-diol, 32, comparable to the premercapturic acid, 33, resulting from the addition of glutathione to 30 by the action of glutathione S-epoxide transferase (32).

While the monocyclic arene oxide, 30, was not sufficiently stable to be isolated from microsomal incubations of benzene, the tricylic naphthalene 1,2-

(35)　　　(34)　　　H_2^*O　　　(36)　　　(37)　　(38)
non-enzymatic
+ GSH

oxide, 34 (which is not in equilibrium with the valence tautomer, 35) one of the three major products, 34, 36, and 37, of microsomal oxidation of napthalene by intact particulate or reconstituted solubilized liver microsomal monooxygenase systems, was isolated as the first arene oxide metabolite (33) more than 20 years after the original postulation (4) of such intermediates in metabolism. Spontaneous aromatization leads almost exclusively to α-naphthol, 37. Stereospecific ring opening by epoxide hydrase leads to the levorotatory *trans*-(IR, 2R)-glycol, 36, $[\alpha]_D$ varying between -50° and -90° (34). If excess glutathione and glutathione trans- ferase are present, the better nucleophile competes suc- cessfully with water and, at the expense of glycol, 36, the premercapturic acid, 38, is formed from the obligatory common labile epoxide intermediate, 34 (33).

The drug metabolizing system, localized in the endoplasmatic reticulum of mammalian liver, even after solubilization and resolution into three components which can be reconstituted (40), still retains two properties: monooxygenase activity (naphthalene → epoxide, 34) and epoxide hydrase activity (epoxide, 34, → glycol, 36) (48). By induction with phenobarbi- tal or with 3-methylcholanthrene animals produce two different purifiable oxygen systems, termed cytochrome P-450 and P-448, respectively (2,5,25). Either system, but more so P-450, even after partial purification, retains epoxide hydrase activity toward naphthalene oxide or other oxides, such as styrene oxide (Table 2),

TABLE 2

Formation of Metabolites from [^{14}C]Naphthalene with Reconstituted Cytochrome

P-450 and P-448 Systems in the Presence of Carrier Naphthalene 1,2-Oxide (48).

	Product (nmoles/nmole hemoprotein/5 min)			
Reconstituted System	Naphthalene Oxide	1-naphthol	Dihydrodiol	Total Metabolites
Cytochrome P-448	1.6	3.6	3.1	8.3
Cytochrome P-450	3.0	5.2	1.9	10.1

suggestive of a close association of the two inducible enzyme systems ("multi-enzyme complex"), closer in P-448 than in P-450 on the basis of trapping experi- ments of ^{14}C-naphthalene oxide in both systems (48).

A reactive arene oxide intermediate open to facile enzymatic and nonenzymatic attack by nucleophiles, such as water, glutathione, mercaptans, must pose a certain danger in metabolism. This danger should increase commensurate with the stability and physiological half-life time of the arene oxide, because this improves the opportunity for attack by nucleophilic groups from the environment. Relatively stable arene oxides are formed from halobenzenes which are agents notorious for causing centrilobular hepatic necrosis especially when the animals, together with, e.g., bromobenzene (5), receive diethyl maleate (42), an agent that siphons off glutathione, or when they are pretreated with phenobarbital which induces P-450. However, pretreatment with 3-methylcholanthrene which induces P-448 offers protection from necrosis by an unknown mechanism and not necessarily by increasing the level of epoxide hydrase (57). Cyclohexene oxide protects rats from necrosis by bromobenzene, possibly by reducing the rate, but not the overall metabolism. These extensive studies have established *a new important function for glutathione* in the liver as a universal protective agent required for drug detoxification (42).

The availability of $1-^2H-$ and $2-^2H-$naphthalene,

(39)   (41)   (43)   (42)   (40)

Rate-determining step

(44)   (45)   (46)
(triangular
transition
state)

39, **40**, and its 1,2-oxides, **41**, **42**, permitted the answer to several questions (3). Firstly, is there an NIH-shift in the "direct" conversion of the two naphthalenes, **39** and **40**, to α-naphthol, **46**, by liver microsomes and are the retentions comparable? Table 3 shows that the answer to this question is in the affirmative. Secondly, are the retentions of deuterium observed in the spontaneous aromatization of the two

naphthalene oxides, 41 and 42, to α-naphthol, 46, comparable? Table 3 shows that under neutral or basic
conditions the retentions of deuterium in α-naphthol
are identical, indicative of the *common intermediate*,
43, the keto tautomer of α-naphthol which can be demonstrated directly *in situ* at -196° by UV irradiation of
naphthalene 1,2-oxide and which rearranges spontaneously

TABLE 3

Hydroxylation of [1-^{2}H]- and [2-^{2}H]Naphthalene with
Liver Microsomes and Two Chemical Model Systems (31).
The Retentions are Accurate to ± 8%.[a]

	% Retention in 1-Naphthol from	
Oxidant	[1-^{2}H]-Naphtha-lene	[2-^{2}H]-Naphtha-lene
Microsomes	64	64
Pyridine N-oxide, hυ (29)	68	64
m-Chloroperoxybenzoic acid (31)	60	68

[a]Ring proton exchange of 1-naphthol with the media does
not occur under these conditions.

and quantitatively to α-naphthol on warming to the
temperature of dry ice. Retentions of deuterium in the
acid-catalyzed isomerizations of oxides 41 and 42 are
dissimilar (Table 4). As a possible intermediate the
triangular transition state, 44, resulting from intramolecular hydride shift was considered and rejected,
because the primary isotope effect expected from a
fully deuterated arene oxide was not observed (37).
The rate determining step in the spontaneous rearrangement is now considered to be the development of a charge
in the zwitter-ionic transition state, 45, without transfer of ^{2}H, while in the acid-catalyzed isomerizations

117

it is the collapse of the protonated arene oxide to the carbonium intermediate.

TABLE 4

Deuterium Retention after Isomerization of [1-^{2}H]- and [2-^{2}H]Naphthalene Oxide to 1-Naphthol (3). The Oxides Contained 1.00 Deuterium Atom, and Retentions are Accurate to $\pm$ 2%.

Isomerization Conditions	% Deuterium Remaining in 1-Naphthol after Isomerization	
	1-Deuterio	2-Deuterio
Liver microsomes (pH 9.0 Tris buffer)	75	72
Acetic acid	70	83
pH 3	59	85
pH 4	58	85
pH 5.5	71	84
pH 7.0	80	81
pH 8.5	81	80

The NIH-shift is not restricted to H, D or T, but occurs also with halogen and alkyl groups. p-Chlorophenylalanine, 47, shows migration of chloride to m-chlorotyrosine, 48, in addition to m-hydroxylation, 49, as well as loss of chlorine, 50 (22). Two of these products, 48, 49, are derivable from a common chlorinated arene oxide intermediate. Likewise, halogenated drugs show halogen migrations both *in vivo* and *in vitro* by the action of non-specific hepatic monooxygenases. Both direct, 54, or vicinal, 53, hydroxylation of a methyl substituent, are observed in the oxidation of methyl-substituted benzenes (38) with microsomes, but

(47)   (48) (17 parts)   (49) (1 part)   (50) (2 parts)

(51)   (52) (11 parts)   (53) (1 part)   (54) (15 parts)

4-methylphenylalanine, 51, shows a unique migration to m-methyltyrosine, 52, by the action of phenylalanine hydroxylase (10).

A model for this methyl migration is provided by p-xylene oxide, 56, which isomerizes to a mixture of unrearranged, 57, and rearranged, 58, xylenols in a

(55)   (56)   (57)   (58)

ratio markedly dependent on pH (Fig. 2) (38). The ratio of phenols, 57,58, in the spontaneous non-enzymatic isomerization at physiological pH (1 part *versus*

119

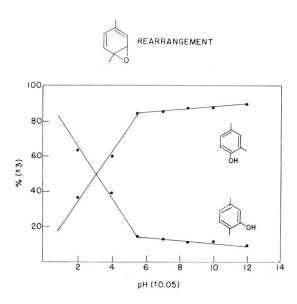

Fig. 2.   Isomerization of 1,4 dimethylbenzene 1,2
oxide to 2,4 dimethylphenol and 2,5 dimethylphenol as a function of pH.   (From ref. 38)

8 parts) compares well with the ratio of 3-hydroxy-4-
methyl-phenylalanine, 53, and m-tyrosine, 52, viz., 1
part versus 11 parts.   Microsomes convert p-xylene, 55,
exclusively to 57, presumably via another oxide, 3,6-
dimenthyl oxide, 56a.

The rearrangement of 1,4-dimentylbenzene oxide, 56,
at pH values < 3.5 leads, with participation of solvent, 59, to cis- and trans-1,4-dimethyl-2,5-cyclohexa-
diene-1,4-diols (60 and 61) which rapidly exchange isotopic oxygen in the presence of $H_2^{18}O$ (56).   This

120

example shows how a monooxygenase could hydroxylate an aromatic substrate without incorporation of molecular oxygen and how weak nucleophiles easily open arene oxides. That such openings may proceed by 1,2-, 63, 1,4-, 64, and 1,6-addition, 65, of nucleophiles was shown by tagging benzene oxide, 62, with deuterium in the 2,5-positions (28).

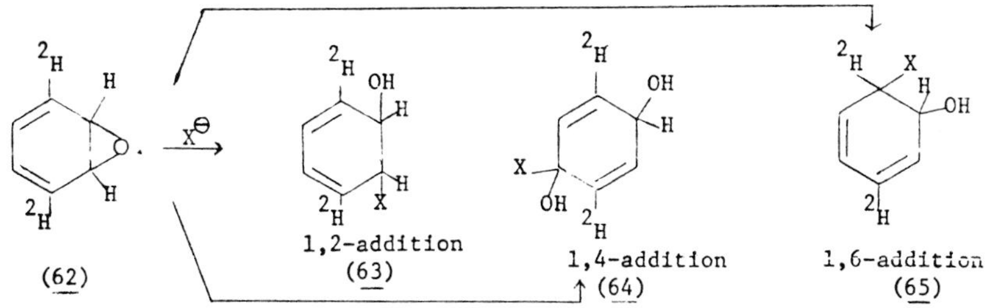

1,2-addition
(63)

1,4-addition
↑ (64)

1,6-addition
(65)

(62)

The isomerization of 8,9-indane oxide, 66, to 4-indanol, 68, rather than proceeding by a dienone →

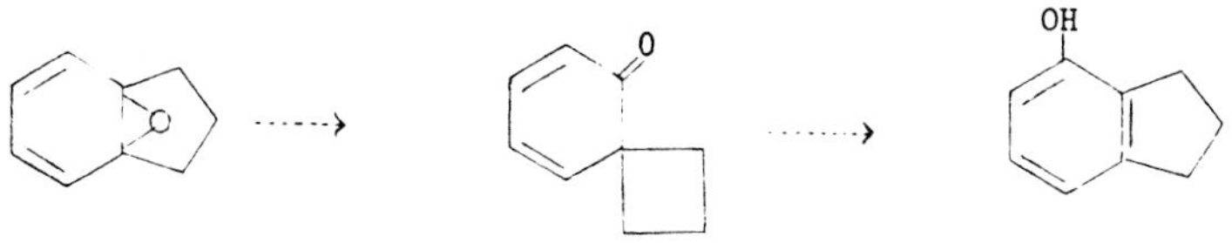

(66)   (67)   (68)

phenol rearrangement (54) or a homoallylic 1,6- addition of water (16), may also involve a circumambulatory oxygen movement (69 → 70 → 71 → 72 → 73-74) explaining the formation of both 4- and 5-indanol (6).

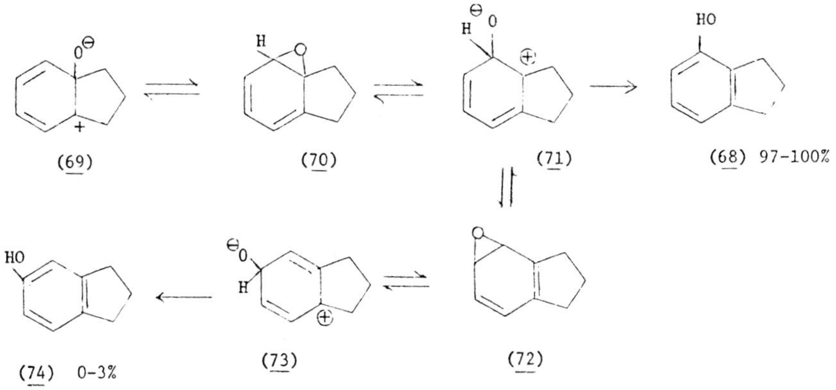

(69)   (70)   (71)   (68) 97-100%

(74) 0-3%   (73)   (72)

121

Polycyclic aromatic hydrocarbons become cancero-
genic, according to new evidence, through metabolism to
their arene oxides (51,20,26).

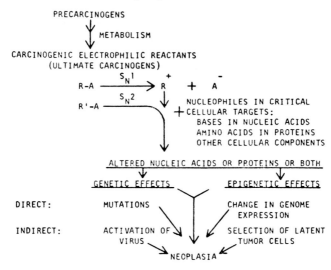

Fig. 3. Some possible mechanisms of carcinogenesis by
the ultimate carcinogenic electrophilic reac-
tants derived from chemical carcinogens or
precarcinogens. (From ref. 41)

As Figure 3 (41) shows, the polycyclic hydrocarbon
is only the *precarcinogen*. In metabolism this becomes
the ultimate carcinogen, the arene oxide, which inter-
acts with cell-bound nucleophiles.

Not only "K-region" oxides, such as 7,12-dimethyl-
benz(a)anthracene-5,6-oxide, 75, but also non K-region

(75)   (76)   (77)

oxides, *e.g.*, of phenanthrene, 76,77, easily accessible
by improved syntheses (19,55), can now be studied enzymatically, physiologically and kinetically and compared
with the non-spontaneous, acid-catalyzed isomerization
of the K-region 9,10-oxide of phenanthrene (36).

The NIH-shift is of considerable diagnostic value:
oxidation systems that do not show the shift (Table 5)
are not valid models for monooxygenases. On the other
hand, oxidants that show the shift, must go through
arene oxides in the same way as the pertinent enzymatic
hydroxylations. Although, understandably, an arene
oxide intermediate could not be demonstrated with tri-
fluoroperacetic acid (31), photolysis of pyridine-N-
oxide in the presence of 1-^{2}H-naphthalene, 78, gave
both naphthalene 1-^{2}H-1,2-oxide, 79, as well as 2-^{2}H-
1-naphthol, 80 (29). The exact nature of the oxidizing

(78)   (79)   (80)

species, formally an electron-deficient oxygen termed
oxene in analogy to *carbene*, remains a challenging subject for future investigations.

Experiments are still continuing to demonstrate
the elusive oxide of phenylalanine. Dihydrophenylala-
nine, 81, is a poor substrate for the hydroxylase and

(81)   (82)   (83)

the products (hypothetical structure, 82) seem to be
ketonic reminiscent of anticapsin, 83, from *streptomyces griseoplinus* which is thought to originate from
shikimic acid rather than from tyrosine (44). A rela-

123

TABLE 5

Differentiation of Oxidants by the Criterion of the NIH-Shift

Model system		R	Retention deuterium or tritium*
FENTON'S, H_2O_2, Fe^{2+}, EDTA	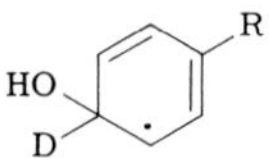	$-NHC(=O)-CH_3$	1.9*
		$-Cl$	2
UDENFRIEND'S, O_2, Fe^{2+}, EDTA, ascorbate or tetrahydropteridine	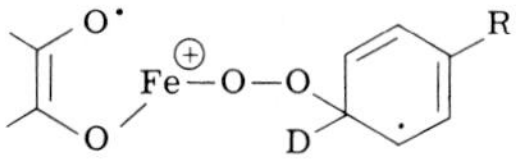	$-NHC(=O)-CH_3$	1.2-1.9*
		$-Cl$	1
H_2O_2, Fe^{2+}, Catechol	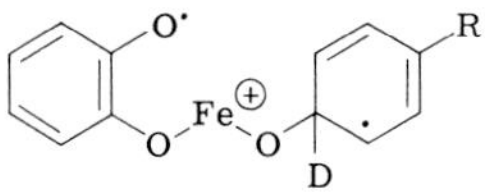	$-NHC(=O)-CH_3$	1*
		$-Cl$	4
Peroxytrifluoroacetic acid	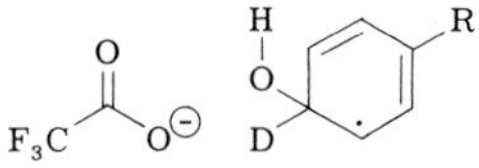	$-NHC(=O)-CH_3$	8
		$-OCH_3$	8
	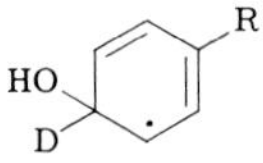	$-Cl$	70
		$-CH_3$	68
Photolysis of pyridine-N-oxide	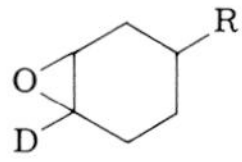	$-OCH_3$	45
Carbenes + oxygen	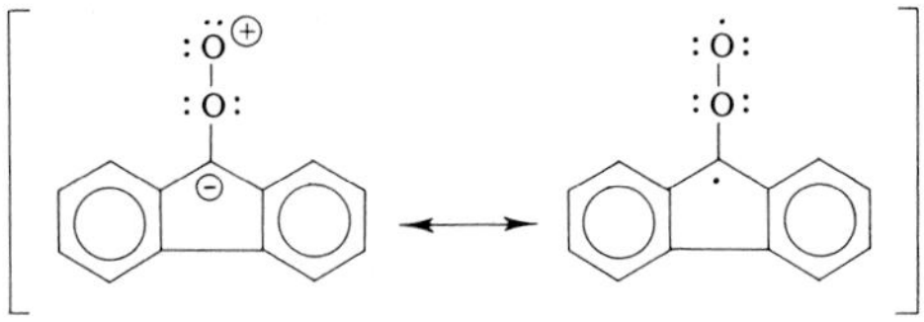 	16% on	
Oxytransition metals	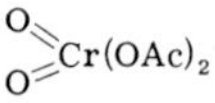 high on 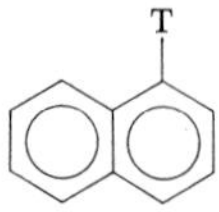		

124

ted 2,3-oxide of phenylalanine, <u>84</u>, has been implied
in the biosynthesis of both gliotoxin, <u>85</u>, and aratonin,
<u>86</u> (35,45). The latter contains an oxepin ring, a type
of metabolite which has so far not been encountered in
mammalian metabolism.

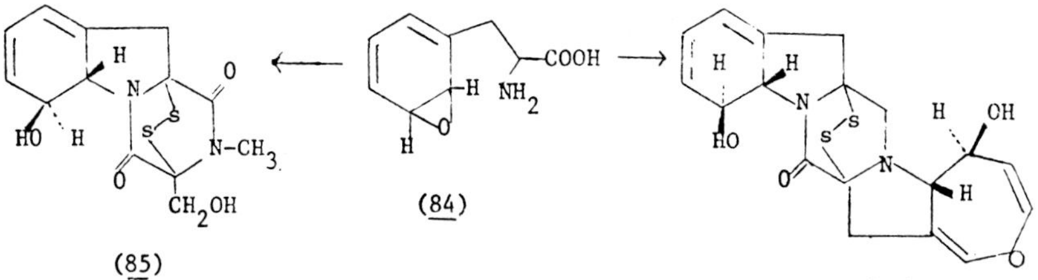

(85) (84) (86)

Modified arene oxides not only cause but also al-
leviate cancer: the plant *Tripterygium wilfordi* contains the triepoxide tripolide, <u>87</u>, which has marked
antileukemic properties (39). The comparable models,

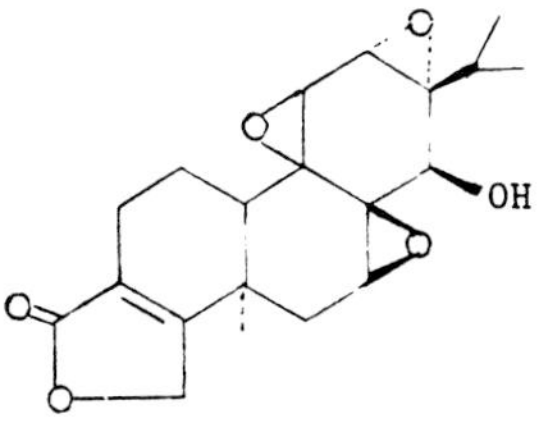

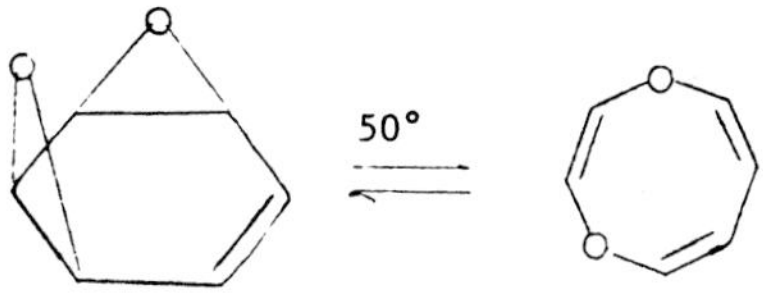

(87) (88) (89)

the quasi-aromatic *syn*-di- and tri-oxides of benzene,
<u>88</u>, <u>90</u>, have recently become available and show valence
tautomerism to 1,4-dioxocin, <u>89</u>, and to *cis-cis-cis*-
1,4,7-trioxacyclononatriene, <u>91</u>, while *anti*benzene

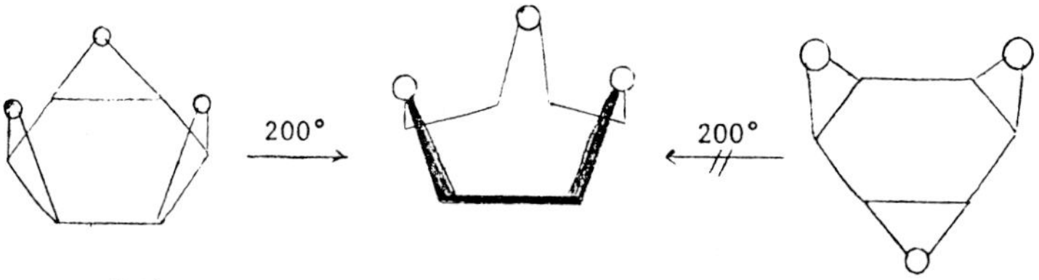

(90) (91) (92)

trioxide, <u>92</u>, is stable to thermal rearrangement (1).

125

The importance of the NIH-shift in the biosynthesis of plant products is a wide field which has been reviewed recently (15).

References

1. Altenbach, H.-J. and Vogel, E. Valence tautomerism of 1,4-dioxocin with *syn*-benzene dioxide. Angew. Chem. Internat. Ed. Engl. <u>11</u>, 937 (1972); Vogel, E., Altenbach, H.-J. and Sommerfeld, C.-D. Valence isomerization of *syn*-benzene trioxide to *cis, cis, cis*-1,4,7-trioxacyclononatriene, *anti*-benzene trioxide. Angew. Chem. Internat. Ed. Engl. <u>11</u>, 939 (1972); Schwesinger, R. and Prinzbach, H. *Trans*-Bis-and-tris-homobenzene derivatives; synthesis of cycloheptatriene-3,4-dicarboxylic acid derivatives. Angew. Chem. Internat. Ed. Engl. <u>11</u>, 940 (1972); Foster, C.H. and Berchtold, G.A. Synthesis of *trans*-benzene trioxide. J. Am. Chem. Soc. <u>94</u>, 7939 (1972).

2. Alvares, A.P., Schilling, G., Levin, W. and Kuntzman, R. Studies on the induction of CO-binding pigments in liver microsomes by phenobarbital and 3-methylcholanthrene. Biochem. Biophys. Res. Commun. <u>29</u>, 521 (1967).

3. Boyd, D.R., Daly, J.W. and Jerina, D.M. Rearrangement of [1-^{2}H]-and [2-^{2}H]naphthalene 1,2-oxides to 1-naphthol. Mechanism of the NIH shift. Biochemistry <u>11</u>, 1961 (1972).

4. Boyland, E. The biological significance of metabolism of polycyclic compounds. Biochem. Soc. Symp. <u>5</u>, 40 (1950).

5. Brodie, B.B., Reid, W.D., Cho, A.K., Sipes, G., Kirshna, G. and Gilette, J.R. Possible mechanism of liver necrosis caused by aromatic organic compounds. Proc. Nat. Acad. Sci. U.S.A. <u>68</u>, 160 (1971).

6. Bruice, P.Y., Kasperek, G.J., Bruice, T.C., Yagi,
    H. and Jerina, D.M. The oxygen walk as a comple-
    mentary observation to the NIH shift. J. Am. Chem.
    Soc. 95, 1673 (1973).

7. Bu'Lock, J.D. and Ryles, A.P. The biosynthesis
    of the fungal toxin gliotoxin; the origin of the
    "extra" hydrogens as established by heavy isotope
    labelling and mass spectrometry. J. Chem. Soc. C,
    Chem. Commun. 1404 (1970).

8. Creveling, C.R. and Daly, J.W. A method for the
    determination of dopamine-β-hydroxylase in crude
    tissue. Pharmacologist 7, 157 (1965).

9. Creveling, C.R. and Daly, J.W. Assay of enzymes
    of catecholamines biosynthesis and metabolism, in
    Methods of Biochemical Analysis, Vol. 19, D. Glick
    (Ed.), John Wiley & Sons, p. 153, 1971.

10. Daly, J. and Guroff, G. Production of m-methyl-
    tyrosine and p-hydroxymethylphenylalanine from
    p-methylphenylalanine by phenylalanine hydroxylase.
    Arch. Biochem. Biophys. 125, 136 (1968).

11. Daly, J., Guroff, G., Udenfriend, S. and Witkop, B.
    Hydroxylation-induced migrations of tritium in
    several substrates of liver aryl hydroxylases.
    Arch. Biochem. Biophys. 122, 218 (1967).

12. Daly, J. and Jerina, D. Migration of deuterium
    during aryl hydroxylation III. Effect of ortho-
    and meta-substituents. Arch. Biochem. Biophys.
    134, 266 (1969).

13. Daly, J., Jerina, D., Farnsworth, J. and Guroff, G.
    The migration of deuterium during aryl hydroxyla-
    tion II. Effect of induction of microsomal hydrox-
    ylases with phenobarbital or polycyclic aromatic
    hydrocarbons. Arch. Biochem. Biophys. 131, 238
    (1969).

14. Daly, J., Jerina, D. and Witkop, B. Migration of deuterium during hydroylation of aromatic substrates by liver microsomes: I. Influence of ring substitutents. Arch. Biochem. Biophys. 128, 517 (1968).

15. Daly, J.W., Jerina, D.M. and Witkop, B. Arene oxides and the NIH shift: the metabolism, toxicity and carcinogenicity of aromatic compounds. Experientia 28, 1129 (1972).

16. Daly, J.W., Jerina, D.M., Ziffer, H., Witkop, B., Klarner, G.F. and Vogel, E. Enzymatic hydration of 8,9-indan oxide. Homoallylic addition of water. J. Am. Chem. Soc. 92, 702 (1970).

17. Ellis, B.E. and Amrhein, N. The NIH-shift during aromatic ortho-hydroxylation in higher plants. Phytochemistry 10, 3069 (1971).

18. Friedman, P.A., Kaufman, S. and Kang, E.S. Nature of the molecular defect in phenylketonuria and hyperphenylalaninaemia. Nature 240, 157 (1972).

19. Goh, S.H. and Harvey, R.G. K-region arene oxides of carcinogenic aromatic hydrocarbons. J. Am. Chem. Soc. 95, 242 (1973).

20. Grover, P.L., Hewer, A. and Sims, P. Epoxides as microsomal metabolites of polycyclic hydrocarbons. FEBS Letters 18, 76 (1971).

21. Guroff, G., Daly, J.W., Jerina, D.M., Renson, J., Witkop, B. and Udenfriend, S. Hydroxylation-induced intramolecular migrations - the NIH shift. Science 158, 1524 (1967).

22. Guroff, G., Kondo, K. and Daly, J. The production of meta-chlorotyrosine from para-chlorophenylalanine by phenylalanine hydroxylase. Biochem. Biophys. Res. Commun. 25, 622 (1966).

23. Guroff, G., Reifsnyder, C.A. and Daly, J.W. Retention of deuterium in p-tyrosine formed enzymatically from p-deuterophenylalanine. Biochem. Biophys. Res. Commun. 24, 720 (1966).

24. Hyaishi, O. Oxygenases, Enzymic Activation of Oxygen. Proceedings of the Robert A. Welch Foundation, Conferences on Chemical Research. XV. Bioorganic Chemistry and Mechanisms, p. 185, Houston, Texas, Nov. 1-3, 1971.

25. Hildebrandt, A., Remmer, H. and Estabrook, R.W. Cytochrome P-450 of liver microsomes--one pigment or many. Biochem. Biophys. Res. Commun. 30, 607 (1968).

26. Huberman, E., Aspiras, L., Heidelberger, C., Grover, P.L. and Sims, P. Mutagenicity to mammalian cells of epoxides and other derivatives of polycyclic hydrocarbons. Proc. Nat. Acad. Sci. U.S.A. 68, 3195 (1971).

27. Hutton, J.J., Tappel, A.L. and Udenfriend, S. Cofactor and substrate requirements of collagen proline hydroxylase. Arch. Biochem. Biophys. 118, 231 (1967).

28. Jeffrey, A.M., Yeh, H., Jerina, D.M., DeMarinis, R.M., Foster, C.H., Piccolo, C.H. and Berchtold, G.A. Stereochemical course in reactions between nucleophiles and arene oxides. In preparation.

29. Jerina, D.M., Boyd, D.R. and Daly, J.W. Photolysis of pyridine-N-oxide: an oxygen atom transfer model for enzymatic oxygenation, arene oxide formation, and the NIH shift. Tetrahedron Letters 457 (1970).

30. Jerina, D.M., Daly, J.W. and Witkop, B. Deuterium migration during the acid-catalyzed dehydration of 6-deuterio-5,6-dihydroxy-3-chloro-1,3-cyclohexadiene,

a nonenzymatic model for the NIH shift. J. Am. Chem. Soc. 89, 5488 (1967).

31. Jerina, D.M., Daly, J.W. and Witkop, B. Migration of substituents during hydroxylation of aromatic substrates (NIH shift). Oxidations with peroxytrifluoroacetic acid. Biochemistry 10, 366 (1971).

32. Jerina, D.M., Daly, J.W., Witkop, B., Zaltzman-Nirenberg, P. and Udenfriend, S. Role of the arene oxide-oxepin system in the metabolism of aromatic substrates I. *In vitro* conversion of benzene oxide to a premercapturic acid and a dihydrodiol. Arch. Biochem. Biophys. 128, 176 (1968).

33. Jerina, D.M., Daly, J.W., Witkop, B., Zaltzman-Nirenberg, P. and Udenfriend, S. 1,2-Naphthalene oxide as an intermediate in the microsomal hydroxylation of naphthalene. Biochemistry 9, 147 (1970).

34. Jerina, D.M., Ziffer, H. and Daly, J.W. The role of the arene oxide-oxepin system in the metabolism of aromatic substrates. IV. Stereochemical considerations of dihydrodiol formation and dehydrogenation. J. Am. Chem. Soc. 92, 1056 (1970).

35. Johns, N. and Kirby, G.S. The biosynthesis of gliotoxin; possible involvement of a phenylalanine epoxide. J. Chem. Soc. C, Chem. Commun. 163 (1971).

36. Kasperek, G.J. and Bruice, T.C. The mechanism of the aromatization of arene oxides. J. Am. Chem. Soc. 94, 198 (1972).

37. Kasperek, G.J., Bruice, T.C., Yagi, H. and Jerina, D.M. Differentiation between the concerted and stepwise mechanisms for aromatization (NIH shift) of arene epoxides. J. Chem. Soc. C, Chem. Commun. 784 (1972).

38. Kaubisch, N., Daly, J.W. and Jerina, D.M. Arene oxides as intermediates in the oxidative metabolism of aromatic compounds. Isomerization of methyl-substituted arene oxides. Biochemistry 11, 3080 (1972).

39. Kupchan, S.M., Court, W.A., Dailey, R.G., Jr., Gilmore, C.J. and Bryan, R.F. Triptolide and tripdiolide, novel antileukemic diterpenoid tri-epoxides from *Tripterygium wilfordii*. J. Am. Chem. Soc. 94, 7194 (1972).

40. Lu, A.Y.H., Kuntzman, R., West, S., Jacobson, M. and Conney, A.H. Reconstituted liver microsomal enzyme system that hydroxylates drugs, other foreign compounds, and endogenous substrates II. Role of the cytochrome P-450 and P-448 fractions in drug and steroid hydroxylations. J. Biol. Chem. 247, 1727 (1972).

41. Miller, J.A. Carcinogenesis by chemicals: an overview - G.H.A. Clowes Memorial Lecture. Cancer Res. 30, 559 (1970).

42. Mitchell, J.R., Jollow, D.J., Potter, W.Z., Gilette, J.R. and Brodie, B. Acetaminophen-induced hepatic necrosis. IV. Protective role of glutathione, in press.

43. Nagatsu, T., Levitt, M. and Udenfriend, S. A rapid and simple radioassay for tyrosine hydroxylase activity. Anal. Biochem. 9, 122 (1964).

44. Neuss, N., Molloy, B.B., Shah, R. and DeLa Higuera, N. The structure of anticapsin, a new biologically active metabolite of *streptomyces griseoplanus*. Biochem. J. 118, 571 (1970).

45. Neuss, N., Nagarajan, R., Molloy, B.B. and Huckstep, L. Aranotin and related metabolites. II. Isolation, characterization, and structures of two new metabolites. Tetrahedron Letters 4467 (1968).

46. Niehaus, W.G. and Schroepfer, G.J. Enzymatic stereospecificity in the hydration of epoxy fatty acids. Stereospecific incorporation of the oxygen of water. J. Am. Chem. Soc. 89, 4227 (1967).

47. Oesch, F. and Daly, J. Solubilization, purification, and properties of a hepatic epoxide hydrase. Biochem. Biophys. Acta 227, 692 (1971).

48. Oesch, F., Jerina, D.M., Daly, J., Lu, A.Y.H., Kuntzman, R. and Conney, A.H. A reconstituted microsomal enzyme system that converts naphthalene to *trans*-1,2-dihydroxy-1,2-dihydronaphthalene *via* naphthalene-1,2-oxide: presence of epoxide hydrase in cytochrome P-450 and P-448 fractions. Arch. Biochem. Biophys. 153, 62 (1972).

49. Reed, D., Vimmerstedt, J., Jerina, D.M. and Daly, J.W. Formation of phenols from aromatic substrates by plant and animal mono-oxygenases: the effect of adjacent deuteriums on the magnitude of the NIH shift of tritium. Arch. Biochem. Biophys. 154, 642 (1973).

50. Renson, J., Daly, J., Weissbach, H., Witkop, B. and Udenfriend, S. Enzymatic conversion of 5-tritio-tryptophan to 4-tritio-5-hydroxytryptophan. Biochem. Biophys. Res. Commun. 25, 504 (1966).

51. Selkirk, J.A., Huberman, E. and Heidelberger, C. An epoxide is an intermediate in the microsomal metabolism of the chemical carcinogen, dibenz(a,h)-anthracene. Biochem. Biophys. Res. Commun. 43, 1010 (1971).

52. Senoh, S., Creveling, C.R., Udenfriend, S. and Witkop, B. Chemical, enzymatic and metabolic studies on the mechanism of oxidation of dopamine. J. Am. Chem. Soc. 81, 6236 (1959).

53. Ulrich, V. Oxygen activation by the iron (II)-2-
     mercaptobenzoic acid complex. A model for micro-
     somal mixed function oxygenases. Z. Naturforsch.
     <u>24b</u>, 699 (1969).

54. Vogel, E. and Günter, H. Benzene oxide-oxepin va-
     lence tautomerism. Angew. Chem. Intern. <u>6</u>, 385
     (1967).

55. Yagi, H. and Jerina, D.M. Synthesis of non-K-
     region arene oxides. J. Am. Chem. Soc. <u>95</u>, 243
     (1973).

56. Yagi, H., Jerina, D.M., Kasperek, G.J. and Bruice,
     T. A novel mechanism for the NIH-shift. Proc. Nat.
     Acad. Sci. U.S.A. <u>69</u>, 1985 ((1972); Kasperek, G.J.,
     Bruice, T.C., Yagi, H., Kaubisch, N. and Jerina,
     D.M. Solvobjic chemistry of 1,4-dimethylbenzene
     oxide. A new and novel mechanism for the NIH-
     shift. J. Am. Chem. Soc. <u>94</u>, 7876 (1972).

57. Zampaglione, N., Jollow, D.J., Mitchell, J.R.,
     Stripp, B. and Gilette, J.R. Role of detoxifying
     enzymes and bromobenzene-induced liver necrosis.
     J. Pharmacol. Exp. Therap., in press.

SEPARATION OF TRANSCRIBABLE AND REPRESSED CHROMATIN

Robert T. Simpson

Section on Developmental Biochemistry
National Institute of Arthritis, Metabolism,
and Digestive Diseases
National Institutes of Health
Bethesda, Maryland 20014

Cell biology has become an area of study with
many ramifications, going beyond the basic understanding of cellular function to more applied, clinical
aspects in diverse diseases. Probably the most fascin-
ating question in modern cell biology concerns the
regulation of gene activity in differentiated cells.
What turns genes on and off in an orderly fashion during the temporal course of development? What allows
for the differentiation of a stem cell into a mature,
functional species, and conversely, what changes take
place to lead to dedifferentiation and neoplasia?

There are probably several types of gene regulation in higher organisms: for example, the quick on
and off synthesis of enzymes in response to hormonal
stimuli; the slower, yet reversible, differentiation of
a quiescent tissue to activity, as in the induction of
lactation in mammary gland; and the stable, adult, differentiated state as contrasted to a multipotential
immature stem cell.

Furthermore, in any of these types of regulation,
the level of regulation may differ. Thus, one could
see apparent differences in gene activity resulting
from changes in the transcription of DNA into RNA,
changes in the processing of RNA in the nucleus, changes
in the rate or degree of translation of mRNA into protein, or changes in proteins *per se*, such as altered
turnover or zymogen activation. Now, within this rather

135

complex matrix of levels and types of regulation, it
seems apparent that some regulation of gene activity
will occur by mechanisms similar to those so elegantly
detailed in bacteria and phage. It seems equally
apparent, on the other hand, that some aspects of regulation in animal cells will differ in kind from those
seen in prokaryotes. This latter supposition is strongly supported by a comparison of the complexity of the
genetic apparatus in prokaryotes, as exemplified by
E. coli, and eukaryotes, with man chosen as example
(Table 1).

TABLE 1

PROKARYOTE	EUKARYOTE
DNA	
0.02×10^{-12} GM /CELL	7×10^{12} GM/CELL
SINGLE COPY	FREQUENT REPEATED SEQUENCES
PROTEIN	
SMALL AMOUNT ACIDIC PROTEIN	HISTONE – I GM/GM DNA
	NON-HISTONE PROTEIN – 0.1-1.5 GM/GM DNA
RNA	
NASCENT mRNA	UP TO 0.15 GM/GM DNA – NASCENT mRNA
	AND OTHER TYPES
CONTROL	
RELATIVELY DISCRETE ALTERATIONS WITH SINGLE STIMULI	SIMPLE STIMULI OFTEN GIVE RISE TO A MULTIPLICITY OF EFFECTS

The DNA of higher organisms is present at 100 to
1000 times the mass per cell of that existing in simpler organisms. Some eukaryotic DNA exists as varying
numbers of repeated sequences, as evidenced by the
studies of Britten and Kohne (5) and others. In contrast to the nearly naked DNA of prokaryotes, the DNA
of eukaryotes is present in a nucleoprotein complex
called chromatin, the diffuse interphase form of the
metaphase chromosome. Chromatin contains a large

amount of proteins. Five main classes of histones, highly basic proteins of low molecular weight, are present in an amount equal to the DNA mass in all higher cell chromatins. Additionally, a highly variable amount of very heterogeneous, acidic, nonhistone protein is associated with the eukaryotic genome. The amount of nonhistone protein varies from about 0.05 to 1.5 times the mass of DNA present. RNA is present with DNA in both groups, a small amount of nascent message in prokaryotes and larger amounts of mRNA, heterogeneous nuclear RNA and, perhaps, specific controlling chromosomal RNA in the higher organisms. As a final difference, the types of control in bacterial systems seem to be relatively discrete, one or more enzymes in a given metabolic pathway responding to a single stimulus. In contrast, animal cells are characterized by complex responses to single external stimuli, as exemplified by the variety of metabolic responses which occur consequent to the stimulation of cells by cortisol or insulin.

Now somehow out of this complexity the cell manages to produce messenger RNA which is appropriate to its particular functions and further, to keep in a repressed state the genes which might be detrimental to the cell. This phenomenon of regulation at the transcriptional level is, in the broadest sense, the direction of my lecture here today. We ask, what in the composition and structure of a given segment of chromatin allows or disallows its translation by RNA polymerase? In the past, a number of indirect approaches have led to some ideas concerning the mechanism of repression of transcription in eukaryotes. Today, I wish to review with you studies from our and others' laboratories which seem to take a more direct approach to the problem. That is, attempts are made to obtain chromatin segments which are transcribable and segments which are repressed. If that is achieved, we can then examine their composition, their structure and their biological functions. After this is accomplished, we can try to relate our conclusions concerning structure-function relationships in the fractionated chromatin to current hypothetical models for animal cell gene regulation. This will constitute the general outline of what I wish to discuss today.

Let me review some of the considerations which suggest that functional and structural heterogeneity do indeed exist in chromatin (Table 2). Total cellular RNA hybridizes at saturation to less than 10% of

TABLE 2

1. LIMITED TRANSCRIPTION, AS MEASURED BY RNA-DNA HYBRIDIZATION

2. TEMPLATE ACTIVITY MUCH LESS THAN DNA

3. CIRCULAR DICHROISM SPECTRUM

4. ELECTRON MICROSCOPIC MORPHOLOGY

5. THERMAL DENATURATION PROFILE

nuclear DNA, suggesting a severe restriction in DNA transcription as a result of the formation of the nucleoprotein complex. When chromatin is compared with protein-free DNA in terms of its template ability, that is, its ability to support DNA-dependent RNA synthesis by RNA polymerase, chromatin DNA serves only about 5-10% as well as template. In addition to this type of evidence suggesting functional heterogeneity, there is ample evidence which also suggests structural heterogeneity. The circular dichroism spectrum of chromatin has been thought by some to result from contributions of two different conformations of DNA (19,23,38). Electron microscopic examination of nuclei or isolated chromatin reveals variable fiber widths, in part related to the method of sample preparation, but, in general, there seems to be agreement that chromatin contains both thicker (100-200 Å) and thinner (30-50 Å) fibers (44; see also 8).

Finally, the thermal denaturation profile of chromatin differs from that of protein free DNA (Figure 1). Chromatin melting is displaced to higher temperatures, reflecting stabilization of the nucleic acid double helix by basic proteins, and strikingly, is

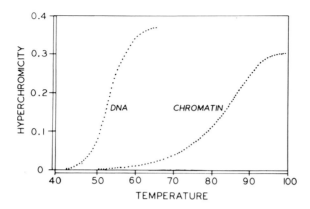

Fig. 1.   Thermal denaturation of DNA and chromatin, as
indicated.   Samples in 0.25 mM EDTA, pH 7.0,
were heated at a rate of 0.25° per min and
absorbance recorded automatically every two
min.   The indicated temperatures are those of
the heating block.

much broader than the cooperative melt of isolated DNA.
It has become apparent that what one observes in the
broad melting of chromatin is a series of overlapping
species, each with its own melting behavior.   By differentiation of these melting curves, Ansevin and coworkers (1) and Li and Bonner (21) have described
three distinct melting regions, suggesting again that
structural heterogeneity exists in unfractionated
chromatin.

     Of particular interest to us as we proceed through
the fractionation experiments and their evaluation is
the presence in chromatin of a small amount of chromatin DNA which melts at quite low temperatures, below
70° in 0.25 mM EDTA, pH 7.0 (Figure 1).   Our interest
in low melting chromatin sequences derives from ingeni-
ous experiments done by McConaughy and McCarthy (24).
They adsorbed chromatin to a hydroxyapatite column
under conditions where double stranded nucleic acid
would bind, but single stranded DNA would not stick to
the column matrix.   A thermal gradient was then applied,
which progressively melted the chromatin on the column

to elute a series of fractions of DNA which corresponded
to various melting ranges of chromatin. Then, using
homologous total cellular RNA, they employed RNA/DNA
hybridization to localize the transcribed sequences in
the elution profile. Most of the transcribable sequences were found in that 3% of the DNA that melted at
the lowest temperature. Hence, we will be interested in
enrichment for low melting sequences as a major criterion
for the separation of repressed and transcribable sequences of chromatin.

Let us depart momentarily from the chemical aspects
of chromatin to look at some of the cytological evidence
concerning repressed and active chromatin in the cell
nucleus. In an early study of RNA synthesis in isolated
thymocyte nuclei, Littau and collaborators (22) were
able to show that RNA was synthesized in areas of the
nucleus that contained relatively diffuse, extended
euchromatin, in contrast to a lack of synthesis in areas
of condensed, granular heterochromatin (Figure 2).
Among others, Zirkin (49) has studied the protein content of inactive and active chromatins by cytochemical
techniques (Figure 3). It is apparent that the diffuse
chromatin of the spermatid and the condensed chromatin
of the mature sperm of *Rana pipiens* both contain histones, stained by alkaline fast green, but only the diffuse, less repressed chromatin of the spermatid contains
significant quantities of nonhistone proteins, as judged
by staining with acid fast green. Similar conclusions
about protein contents have been reached by chemical
analyses of chromatins from tissues with varying metabolic activity. The content of histones seems almost
invariant in a wide variety of tissues, while their
nonhistone content varies from almost none in totally
turned-off tissues such as mature avian erythrocyte to
high values in chromatin from tissues actively engaged
in RNA synthesis, such as liver or tumor cells (2).
Based on these cytologic observations, we therefore
expect to find transcribable chromatin in a more extended conformation than repressed and to find an enrichment of nonhistone proteins in the chromatin segments which are transcribable.

140

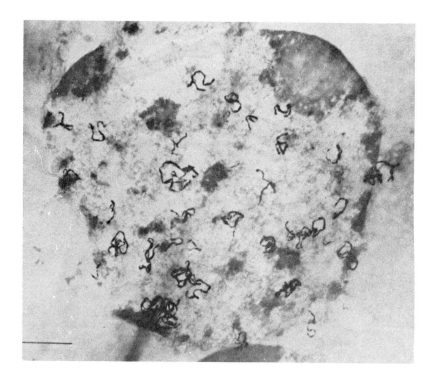

Fig. 2. Electron microscopic autoradiography of RNA
        synthesis in isolated thymocyte nuclei after
        incubation with tritiated uridine. From
        Littau, *et al*. (22).

Frenster (13) has shown continuity between extended and condensed chromatin DNA strands. Hence, it
is necessary to break up the linear structure of the
chromatin fibril in some fashion prior to attempts to
fractionate different species. A diagramatic representation of two models for chromatin structure (Figure 4)
leads to an important conceptual feature for considering chromatin fractionation methods. In either the
Paul (32) or the Crick (9) model, transcribable DNA is
thought to exist as extended regions, while repressed
chromatin DNA is thought to occur as either condensed
globular or supercoiled regions. In order to separate
these two entities it is obviously necessary to reduce

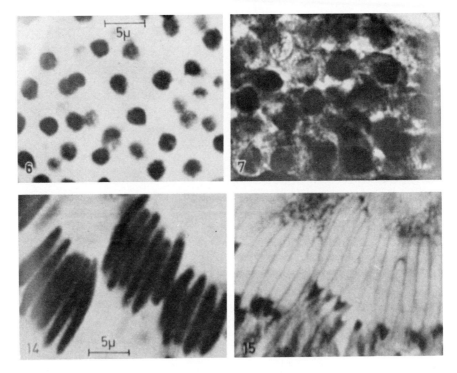

Fig. 3. Cytochemical analysis of protein content in
condensed and extended chromatins. Spermatids
(above) and mature sperm (below) of *Rana
pipiens* were stained with alkaline fast green
to detect histones (left) and with acid fast
green to detect nonhistone proteins (right).
From Zirkin (49).

the size of the particle, hopefully to something about
or less than the size of a transcribed region. Most
commonly this is done by either sonication or mechani-
cal shearing in a blender. Either of these processes
is likely to produce essentially random breaks and consequently generate a spectrum of types of molecules,
varying from pure repressed segments to a small amount
of pure transcribable segments but with most of the
material containing both types of sequences, albeit in
varying proportions. Hence, any method, such as that
early described by Frenster and coworkers (14), that

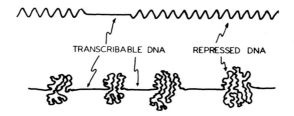

TRANSCRIBABLE DNA   REPRESSED DNA

Fig. 4. Gross representation of models for the struc‐
ture of the interphase chromosome. The models
of Paul (32) and Crick (9) are shown above and
below respectively.

simply fractionates chromatin into two or three frac‐
tions is liable to overlook many features of subfrac‐
tion structure and composition and further is prone to
yield quite variable results. What is required and
what I will discuss today are methods which allow for
a continuous display of molecular types.

Fractionation Methods

The three methods which I will consider for chroma‐
tin fractionation are logical outgrowths, at least ret‐
rospectively, of the cytological characteristics of
putative transcribable and repressed chromatin that I
have already discussed. Thus, since transcribable
chromatin is thought to be enriched in its content of
acidic, nonhistone proteins, one should be able to
achieve a fractionation by ion-exchange chromatography.
This is a method which has been developed in our lab‐
oratory by Dr. Gerald Reeck (35), using ECTHAM-cellu‐
lose as an adsorbent. The extended nature of trans‐
cribable chromatin and the condensed nature of re‐
pressed segments should allow their separation by size-
shape criteria also. Sedimentation on sucrose grad‐
ients has been employed by Chalkley and Jensen (6),
Duerksen and McCarthy (11), Slayter, et al. (44),
Nishiura (26) and Murphy and coworkers (25). Given
equal particle masses, one would expect the condensed
(repressed) chromatin to sediment more rapidly than the

extended (active) chromatin. Finally, an extended molecule will have a greater Stokes radius than a condensed
molecule of the same mass and hence these two types of
chromatin segments should be separable on the basis of
their behavior on gel filtration. A method for chromatin fractionation based on this principle has been developed by Janowski, Nasser and McCarthy (18) and will
be considered, although less extensively than the ion
exchange and sedimentation techniques.

The sedimentation methods are improvements of the
original Frenster, *et al.* (14) technique for fractionation of chromatin - important modifications in that they
allow the display of a spectrum of molecular types.
Sedimentation of sheared myeloma cell chromatin on a
0.17-1.7 M sucrose greadient in 0.01 M Tris·Cl, pH 8.0,
for 16 hours at 22,500 rpm leads to the distribution of
chromatin shown in Figure 5 (25). About 25% of the

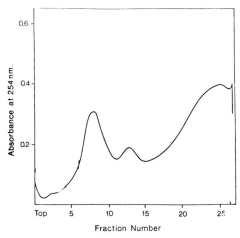

Fig. 5. Fractionation of mouse myeloma chromatin by
         sucrose density gradient sedimentation. Three
         ml of sheared myeloma chromatin was overlaid
         on a thirty ml linear 0.17 - 1.7 M sucrose
         gradient in 0.01 M Tris·Cl, pH 8.0 and cen-
         trifuged for 16 hours at 22,500 rpm and 4° in
         a Spinco SW25 rotor. From Murphy, *et al.*
         (25).

chromatin is pelleted under these conditions and the remainder is distributed as a sharp, very slowly sedimenting peak, amounting to 12-20% of the total, and an intermediate region of rather larger sedimentation constant. By varying conditions of shearing, Murphy *et al.* (25) felt that they could demonstrate that 10-20% of the total chromatin DNA was associated with proteins in a manner that produces an extended, active slowly-sedimenting conformation. Duerksen and McCarthy (11) sedimented sheared mouse hepatoma and crab testis chromatins on steep sucrose gradients for shorter periods of time and similarly obtained a partial resolution of rapidly and slowly sedimenting samples, the rapidly sedimenting species comprising somewhat more than half the total material. Using a tissue thought to be less active in RNA synthesis, Chalkley and Jensen (6) fractionated sheared calf thymus chromatin on shallow sucrose gradients, 5-30% (w/v), and observed a broad, skewed peak containing an envelope of chromatin particles with sedimentation constants varying from 30 S to greater than 130 S.

The functional properties of chromatins isolated by gradient sedimentation have been investigated primarily by assay of the template activity (under conditions of enzyme saturation) of the fractions for RNA synthesis by DNA-dependent RNA polymerase of *E. coli* or, more recently, of homologous animal cell origin. Chalkley and Jensen (6) initially noted that the more slowly sedimenting species of calf thymus chromatin were slightly more transcribable by bacterial RNA polymerase than the rapidly sedimenting chromatin segments, although this tissue is thought to possess a chromatin with quite low *in vivo* template activity.

More recently, Murphy and coworkers (25) have studied the template activity of their myeloma chromatin fractions (Figure 6). The slowly sedimenting peak has a high template activity, nearly 50% that of protein-free DNA on its centripetal edge. Template activity then declines steadily through the intermediate sedimenting species, becoming minimal in the most rapidly sedimenting chromatin segments. The overall

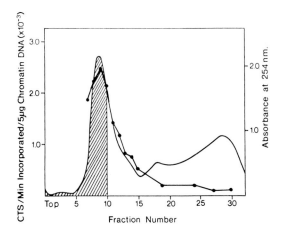

Fig. 6. Template activity of mouse myeloma chromatin
fractionated by sucrose gradient centrifugation. Chromatin was fractionated as in the
legend to Figure 5 and template activity for
E. coli RNA polymerase determined (●——●).
From Murphy, *et al.* (25).

ratio of template activity for the slowest sedimenting,
highest activity fractions to the most rapidly sedimenting, lowest activity fractions is in the range of
15-20. Moreover, the slow, shaded peak, containing
10-20% of the chromatin DNA, has over 80% of the total
template activity of the unfractionated chromatin.
While the illustrated results were obtained with *E.
coli* RNA polymerase, similar differences in template
activity were observed for these fractions when either
the nucleolar or nucleoplasmic homologous RNA polymerase was utilized. Thus, sedimentation in sucrose
density gradients seems to allow the separation of
chromatin fragments which differ in their biological
activity as judged by template capacity. Furthermore
the prediction that the slowly sedimenting species
should be transcribable while the condensed rapidly
sedimenting species should be relatively repressed has
been borne out.
        Janowski, *et al.* (18) have fractionated sheared
chromatin from a variety of sources by gel filtration

on columns of agarose. The solvent chosen for their
studies, 0.15 KCl, 0.1 M MgCl$_2$, 0.001 M mercaptoethanol,
0.01 M Tris·HCl, pH 8.0, is one that may quite possibly
lead to dissociation or at least significant exchange
of the proteins of chromatin (*c.f.* Clark and Felsenfeld
(7)) and consequently caution must be used in inter-
preting their results. Nevertheless, this group has
done some types of experiments with fractionated chromatin that have not been reported by anyone else and
therefore we will consider their results, although with
slight reservations. When chromatin is filtered on
Bio-Rad Agarose A-50m, an initial breakthrough peak
emerges at the void volume of the column, followed by
a broad peak in the fractionating volume of the column
eluate (Figure 7). I remind you that we would expect

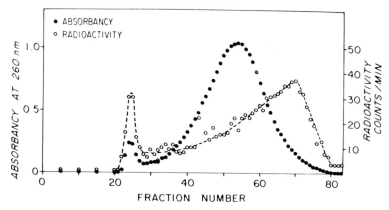

Fig. 7.   Fractionation of mouse liver chromatin by gel
          filtration on Bio-Gel A-50m. Sheared chromatin
          was chromatographed in the solvent listed in
          the text after *in vivo* incorporation of triti-
          ated uridine to label nascent RNA. Absorbance
          and radioactivity are plotted as indicated.
          From Janowski, *et al.* (18).

the initial peak to be the transcribable chromatin fragments, with the repressed, condensed material eluting
at a later volume. Indeed, the results confirm this
supposition. Using mouse liver chromatin from animals

labelled by intraperitoneal injection of tritiated uridine, the results shown in Figure 7 were obtained. Labeled RNA appears at either the position of the initial peak or after the larger optical density peak. The authors interpret their results to indicate that much of the nascent messenger RNA in chromatin is dissociated or degraded during preparation and fractionation, leading to the peak at the end of the elution profile, while the nascent mRNA that remains *in situ* is associated solely with the initial breakthrough peak. Janowski, *et al.* (18) obtained essentially identical results when the distribution of labeled RNA synthesized *in vitro* with *E. coli* RNA polymerase was evaluated.

The final methodological approach to chromatin fractionation which we will consider is the use of ion exchange chromatography on ECTHAM-cellulose columns (35). This method, like the others, allows the display of a continuum of nucleoprotein species but, unlike the others, is the only method currently in use that does not entirely depend on size-shape fractionation. While I am sure that most of you are familiar with the properties of several of the ion exchange cellulosic adsorbents, let me compare the features of the relatively obscure ECTHAM-cellulose with those of DEAE-cellulose, the commonest cationic adsorbent (Table 3). ECTHAM-cellulose

TABLE 3

	DEAE	ECTHAM
GROUP	$-O-CH_2-CH_2-N\begin{smallmatrix}C_2H_5\\C_2H_5\end{smallmatrix}$	$-CHOH-\begin{smallmatrix}CH_2OH\\C-NH_2\\CH_2OH\end{smallmatrix}$
pK	9.5	7.2
CAPACITY	1.0	0.1

has tris(hydroxymethyl) aminomethane as its ionizing group, hence a lower pK than DEAE-cellulose. Further, the capacity of the two celluloses differ by a factor of ten. ECTHAM-cellulose was initially synthesized by Peterson and Kuff (33) for the chromatography of large polyanions, such as nucleoproteins, and was employed by these authors for fractionation of ribosomes. The low pK of the ionizing group and its low capacity allowed the elution of even large polynucleotides under conditions which are relatively mild, in contrast to the high salt concentrations or extremes of pH necessary for elution of most nucleic acids from the commoner cationic celluloses.

We normally fractionate chromatin by loading an ECTHAM-cellulose column equilibrated with 0.01 M Tris·Cl, pH 6.5, to 80% of its capacity with sonicated chromatin, washing briefly with the starting buffer, and then eluting by titrating the binding groups on the column with 0.01 M Tris base containing 0.01 M NaCl. As the basic solution titrates the binding groups on the adsorbent, the chromatin particles are eluted, the weakest bound first, followed by a gradual transition to the tightest bound molecules eluting at the tail of the peak where the pH rises, indicating the completion of the titration of the binding groups on the adsorbent (Figure 8).

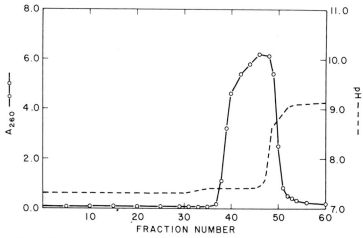

Fig. 8. Chromatography of sonicated rabbit liver chromatin on ECTHAM-cellulose. A 2 gm ECTHAM-cellulose column was loaded with 70 A_{260} units of chromatin in 0.01 M Tris·Cl, pH 7.3, and eluted with 0.01 M Tris – 0.01 M NaCl. Absorbance (———) and pH (----) are plotted. From Reeck, et al. (36).

I outlined previously the presumed significance of low melting sequences in chromatin. The primary criterion we have used in the evaluation of fractionation schemes was enrichment in (or depletion of) low melting sequences in the various fractions. The early eluted chromatin fractions from the ECTHAM-cellulose column melt at significantly higher temperatures than unfractionated chromatin and, more notably, totally lack any segments which melt at temperatures of less than 70° (Figure 9). As the column is developed, the proportion

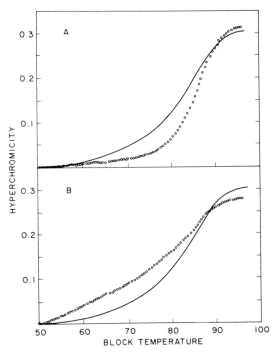

Fig. 9. Thermal denaturation of fractions of rabbit liver chromatin obtained by ECTHAM-cellulose chromatography. After chromatography as in the legend to Figure 8 and dialysis into 0.25 mM EDTA, pH 7.0, early eluted (A) and late eluted (B) fractions were melted as in the legend to Figure 1. The solid lines are the melting profile of unfractionated chromatin. From Reeck, *et al.* (36).

of low melting material gradually incrases, becoming similar to that of unfractionated chromatin 80% through the elution profile for rabbit liver chromatin. Continuing elution brings off fractions with increasing low melting material until at the tail of the peak, a maximal enrichment of about 40-50% of the chromatin segments melting at temperatures below 75° is achieved (Figure 9). Again, it is pleasing to note that the low melting chromatin, which the McConaughy-McCarthy experiments imply can be equated with transcribable chromatin, elutes later than the putative repressed chromatin, consistent with either decreased histone or increased non-histone content in the transcribable regions.

The functional criterion we have applied to this fractionation system is similar to that employed by Janowski and coworkers. Thus, rats were labeled for periods from 10 to 240 minutes by intraperitoneal injection of tritiated orotic acid. Liver chromatin was then isolated, containing all the label as RNA, and fractionated on the ECTHAM-cellulose column. The results indicate that the late eluted, low melting chromatin is indeed associated with more labeled nascent RNA than the early eluted high melting chromatin segments (35). When the ratio of labeled RNA to DNA is plotted, there is a gradual increase from relatively low values at the front of the peak to higher ones through the main portion of the eluted peak (Figure 10). At the end of the elution profile, in the region where we see a maximal augmentation in the proportion of low melting sequences, there is a very rapid increase in the ratio, attaining a limit of about four times the value for the earliest eluted chromatin.

An additional functional measure has recently been described by Janowski, et al. (18). In hormone responsive tissues, such as oviduct or uterus or the liver of adrenalectomized animals, cytoplasmic hormone receptor proteins appear to exist (29,45,46). Hormone bound to these receptors can be shown to interact with chromatin, presumably as a part of the mechanism of gene activation which must accompany the hormone dependent differentiation of the tissue. Incubation of monkey endometrial tissue with labeled estradiol, followed by isola-

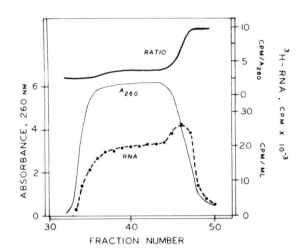

Fig. 10. Distribution of nascent chromatin bound RNA
in rat liver chromatin fractionated on ECTHAM-
cellulose. Rats were labeled with 0.5 mC
tritiated orotic acid for 10 min and chromatin
isolated and chromatographed as in the legend
to Figure 8. Absorbance, radioactive RNA and
their ratio are plotted as indicated.

tion of chromatin and its fractionation in the gel filtration system led to the results shown in Figure 11.
A peak of radioactivity due to estradiol emerges with
the breakthrough peak, that portion of the chromatin
which is thought to be transcribable. There is little
hormone associated with the later eluted chromatin segments but a second peak of hormone does emerge at the
position of free estradiol, thought to be due to dissociation of the hormone during experimental manipula-
tions. Association of hormone with the transcribable
chromatin segments strengthens the contention that
these are indeed the segments of chromatin which are
active in RNA synthesis and further suggests that the
control loci, to which the hormone—receptor protein
complex presumably binds, are closely linked in the intact genome to structural genes which are controlled by
hormone activation.

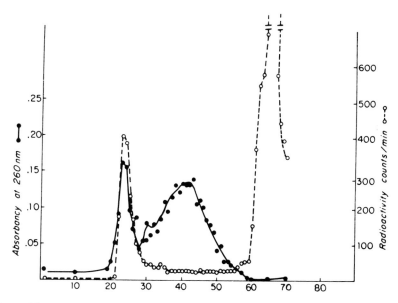

Fig. 11.   Distribution of labeled estradiol in chromatin
           fractionated on Bio-Gel A-50m.   Endometrial
           tissue from a monkey was incubated *in vitro*
           with labeled estradiol for 45 min at 37°.
           Chromatin was isolated and fractionated as in
           the legend to Figure 7.   Absorbance (●) and
           the distribution of labeled hormone (o) are
           plotted.   From Janowski, *et al.* (18).

     An important caveat exists for all the separation
methodologies thus far detailed.   No single method has
absolutely demonstrated that it indeed separated transcribed and repressed chromatin.   This would require,
in addition to template activities with homologous,
nucleoplasmic RNA polymerase and knowledge of the distribution of nascent mRNA, the demonstration by RNA/DNA
hybridization of the localization of transcribed segments of the chromatin.   All this information is just
not available for any single fractionation system.
Nevertheless, the available evidence obtained by a number of fractionation methods on different cellular systems and evaluated by different functional measures is

thus far totally self-consistent. With the caveat in mind, I will proceed to discuss the composition and structure of fractionated chromatin. For ease in purposes of description, we call the extremes of the spectrum of chromatin types *transcribable* and *repressed*.

Composition of Chromatin Fractions

In the examination of the composition and structure of these chromatin types I will, in a few cases, look at the distribution of species throughout the fractionation profile, but, in general, I will compare the properties of the extremes of the spectrum of nucleoprotein types - *i.e.*, slowly and rapidly sedimenting chromatin in the sucrose gradient systems and the early and late eluted chromatin fractions from ECTHAM-cellulose chromatography. Again, the matrix of compositional and structural analysis *vs* type of separation is not complete and I will have to draw selected examples for each parameter examined. First, let us look at the composition of repressed and transcribed chromatins.

When DNA is isolated from fractionated chromatin there seems to be only one feature that differs consistently. Animal cells contain variable amounts of so-called satellite DNA, either GC or AT rich nucleic acid, which consequently bands at a different density from the major portion of the cellular DNA in CsCl or Cs_2SO_4 gradients (48). Studies of Yunis and Yasmineh (48) and Duerksen and McCarthy (11) have established the preferential localization of satellite DNA in the repressed, heterochromatic fractions of chromatin isolated either by differential sedimentation or sucrose gradient sedimentation (Figure 12).

Other studies have localized satellite DNA in the constitutive heterochromatin of animal cells (8,48), a cytological form of chromatin which is thought to be essentially inactive in RNA synthesis throughout the cell cycle. A second type of heterochromatin is that present in nucleoli, which, while condensed in structure, is nevertheless active in synthesis of ribosomal RNA. This constitutes a special case which is confus-

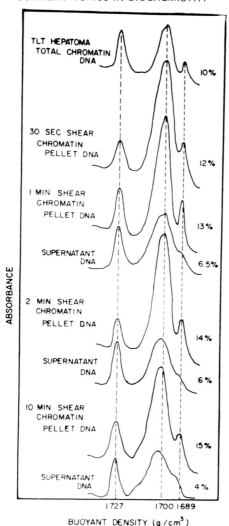

Fig. 12. Analytical pycnography of DNA isolated from chromatin fractions of mouse hepatoma. Chromatin was sheared for varying periods of time, fractionated by differential centrifugation at 3000 xg for 15 min and the DNA isolated from supernatant and pellet analyzed by CsCl density gradient centrifugation. The band at 1.727 is *Myxococcus xanthus* DNA, used as a reference. The % numbers refer to the content of satellite estimated by planimetry. From Duerksen and McCarthy (11).

ing in terms of the generalizations I wish to make and
hence I will ignore it from here on.

Other than a preferential enrichment of satellite
DNA in the repressed chromatin segments, the DNA of all
chromatin fractions appears to be similar. For both
liver and thymus, there is no significant variation in
the melting temperature (hence base composition) or
circular dichroism spectrum of the DNA isolated from
transcribed and repressed chromatin fractions separated
by ECTHAM-cellulose chromatography (Table 4) (34). The

TABLE 4

	CALF THYMUS		RABBIT LIVER	
	EARLY	LATE	EARLY	LATE
T_M	85°	85°	85°	85°
$[\theta]_{275}$	9500	9500	9500	9500
$S_{20,W}$	8.2	7.3	8.8	7.7
MW x 10^{-5}	4.1	2.3	4.8	3.4

sedimentation constant and molecular weight of the DNA
obtained from transcribable chromatin fractions are
slightly lower than those of DNA from repressed chromatin fractions (Table 4). Similar results for melting
of DNA from chromatin fractions were obtained by Duerksen and McCarthy (11) and for molecular size of DNA by
Chalkley and Jensen (6) for chromatins fractionated by
sucrose gradient centrifugation.

In contrast to this general lack of variability
in DNA composition, there is general agreement that
repressed and transcribable chromatin fractions differ
markedly in their protein composition. We have measured
protein content across the elution profile from ECTHAM-
cellulose columns for both rabbit liver and calf thymus
chromatins (42). The content of histone, expressed as
a mass ratio of protein to DNA, of both liver and thymus chromatin is about 1.0-1.05 gm/gm DNA. In the
early part of the elution profile for thymus chromatin,
the histone content is slightly higher than the input
(Figure 13). Coincident with an enrichment for low

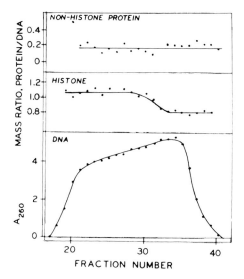

Fig. 13. Distribution of proteins in fractionated calf thymus chromatin. Calf thymus chromatin was fractionated at a twenty-fold scale up from the conditions in the legend to Figure 8, using ECTHAM-cellulose. Total protein was determined by the Lowry method, histone by a Reinecke salt method, and nonhistone protein by difference. The protein contents are plotted as mass ratios of protein to DNA.

melting sequences, about 60% of the way through the thymus chromatin elution profile, there is a decrease in histone content to approach a level of 0.8 gm/gm DNA (Figure 13). In contrast to this change in histone content, the content of nonhistone protein in thymus chromatin is low, about 0.2 gm/gm DNA, and apparently invariant across the elution profile (Figure 13). Chalkley and Jensen (6) agree with these findings in their observation of invariant nonhistone and slightly decreased histone content in slowly sedimenting calf thymus chromatin.

For liver, a simlar situation pertains regarding the distribution of histones in the chromatin fractions. A high initial content of 1.2 gm/gm DNA is

157

decreased for most of the chromatogram to about equal
to the input content. Then again coincident with enrichment for low melting sequences, there is a decrease
to a level of 0.75 gm histone/gm DNA (Figure 14).
Liver, thought to be a more active tissue in RNA synthesis, is characterized by a higher nonhistone protein than thymus, usually about 1 gm/gm DNA. Throughout the major portion of the elution profile the content of nonhistones is constant and about equal to the
input content. Strikingly, at the end of the eluted
peak, even slightly later than the decrease in histone
content, there is a marked increase in the content of
nonhistone proteins, approaching a limit of 2.5 times
DNA mass in the last few fractions at the tail of the
peak (Figure 14). A number of other methods which I

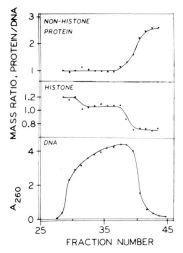

Fig. 14. Distribution of proteins in ECTHAM-cellulose
          fractionated rabbit liver chromatin. Deter-
          minations and method of plotting are as in
          the legend to Figure 13.

will not have time to consider have also suggested
that transcribable chromatin is increased in nonhistone
protein content and perhaps slightly decreased in histone content.

We can now inquire as to the nature of the histones
and nonhistones present in transcribed and repressed
chromatin fractions. To examine the types of histones
present, we use electrophoresis on 25 cm long, acidic,
urea-containing polyacrylamide gels, as initially described by Panyim and Chalkley (30). These are high
resolution gels, allowing the detection of each of the
five main histone classes and, even, some of their
modified variants.

The complexity and confused nomenclature of the
histones is beyond the scope of what we have to say
today and has been reviewed recently (10,17). Suffice
it to say that there are five classes of histones, all
highly basic proteins containing 25-30 mole% lysine +
arginine, with molecular weights ranging from 11,000
to 25,000. In the gel system we use (Figure 15) the
lysine-rich histone, f1, migrates with the lowest mobility, followed by the sulfhydryl-containing histone,
f3, and the three others. Whole chromatin contains
approximately equal amounts of these five proteins.
The fine structure evident for several of the peaks
arises partly from post-synthetic modifications of the
proteins, such as acetylation or phosphorylation, and
partly from primary sequence variants for f1 histone.
Comparing the histones from early and late eluted rabbit liver chromatin (Figure 15) it is apparent that
there is near identity between the electrophoretic patterns of the proteins in transcribed and repressed
chromatin fractions. Only in the case of the content
of the lysine-rich histone, f1, is a difference detectable. The amount of this histone in the latest eluted
fractions from ECTHAM-cellulose columns is about 25%
of its content in unfractionated chromatin. The content of the other histones and, in particular, the relative amounts of the modified variants is not different
for the various fractions of chromatin obtained by
ECTHAM-cellulose chromatography. This latter observation is of interest since some workers have postulated
that histone acetylation or phosphorylation was a critical feature in the creation of transcribably regions
in animal cell chromatin (8,41). Our data do not show

159

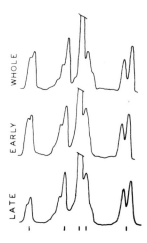

Fig. 15.   Polyacrylamide gel electrophoresis of the
histones of rabbit liver chromatin. Histones
were acid extracted and electrophoresed in
the Panyim and Chalkley (30) system. After
staining with fast green and destaining, the
gels were scanned in an E-C Co. densitometer.
Unfractionated chromatin and early and late
eluted fractions from ECTHAM-cellulose chromatography are shown as indicated. The direction of migration is from left to right and
only about the bottom 25% of the gel scan is
shown. The index marks from left to right
indicate the positions of migration of histones f1, f3, f2b, f2a2 and f2a1. The ordin-
ant is linear with optical density.

any accumulation of modified histones in either repressed
or transcribed chromatin and hence can not provide any
experimental support for this hypothesis.

   A similar situation is usually observed for calf
thymus chromatin histones (Figure 16). Although chromatin from this tissue is thought to be less transcribable than that of liver, a decrease in histone content is
apparent at the end of the elution profile (Figure 13)
and, similar to liver, the decrease in histone content
seems to be characterized by a diminution only in the

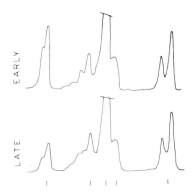

Fig. 16. Gel electrophoresis of the histones from calf thymus ECTHAM-cellulose fractions. Conditions as in the legend to Figure 15.

content of the lysine-rich histone (Figure 16). In addition to the constancy of the amounts of the other histones, there is again no indication of accumulation of modified histones at either end of the elution profile.

On occasion, we obtain preparations of thymus chromatin which contain a significant amount of the dimer of histone f3, the only histone to contain a sulfhydryl group and hence the only one capable of forming a covalently disulfide linked dimeric molecule. We are not at all certain whether the correct *in vivo* case is the presence or absence of dimer. It is of interest, however, that when f3 dimer is present it is confined to the early eluted fractions from ECTHAM-cellulose chromatography (Figure 17), *i.e.*, the repressed and condensed fractions. Coincident with the appearance of low melting material and extended sequences, f3 begins to appear as a monomer, the total amount of the histone remaining constant (Figure 17). Whatever its cause, the presence of f3 dimer solely in condensed chromatin is of interest since this dimeric histone could conceivable crosslink DNA strands to lead to condensation of the nucleoprotein. In this vein, it is of high interest that Bradbury and collaborators (4) have recently postulated a similar role for f1 histone in crosslinking DNA in chromatin into more highly condensed structures. The

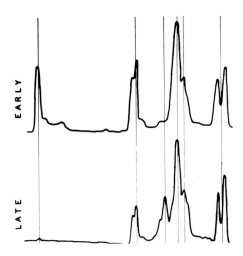

Fig. 17. Gel electrophoresis of the histones from
chromatin fractions isolated from a sample of
calf thymus that contained the dimer of histone f3.   Conditions as in the legend to Figure 15 except that a greater portion of the
gel scan is shown and the leftmost index mark
indicates the position of migration of the f3
dimer.

two forms of histone that have been postulated to induce formation of condensed structures are thus present
in repressed and absent or diminished in transcribed
chromatins, consistent with the postulated physical
properties of these two species.

The nonhistone proteins of metabolically active
tissues, such as liver, are much more diverse than the
histones. By electrophoresis on sodium dodecyl sulfate
containing gels at least 25-30 distinct bands are
usually seen (12) and the possible charge heterogeneity
within this size-based separation is unknown. Superimposed on a background of overall similarity, there
are numerous quantitative and qualitative differences
between the nonhistone proteins of repressed and transcribable chromatins (42). For example in rabbit liver
chromatin fractionated by ECTHAM-cellulose chromatography, repressed chromatin is seen to contain a group of

lower molecular weight nonhistones which are absent in transcribable fractions (Figure 18). A marked increase

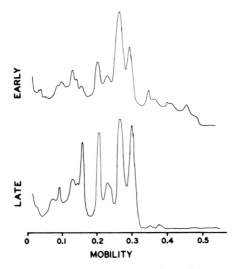

Fig. 18. Densitometric scans of sodium dodecyl sulfate polyacrylamide gels of the nonhistone proteins from early and late eluted ECTHAM-cellulose fractions of rabbit liver chromatin. Samples were obtained by SDS extraction of dehistonized chromatin fractions and electrophoresed on 15 cm long SDS gels. Mobilities are relative to the migration velocity of the bromphenol blue tracking dye.

in the content of the nonhistone protein with mobility 0.16 in transcribed chromatin has been noted previously (36). Further, the proteins with mobilities of 0.21 and 0.31 appear to be present in greatly enhanced amounts in transcribed chromatin *vs* repressed (Figure 18).

The repressed and transcribable fractions of myeloma chromatin fractionated by sucrose gradient sedimentation also differ in nonhistone protein content (25). In this case, however, the higher molecular weight nonhistone proteins that differ between the two samples seem to be preferentially localized in the repressed

fractions, while the active fractions are characterized
by an augmentation in their content of relatively
smaller nonhistone proteins (Figure 19). Different

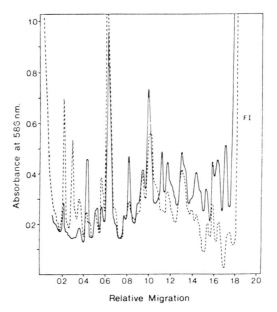

Fig. 19. Densitometric scans of the nonhistone proteins
from slowly (——) and rapidly (----) sedimenting fractions of mouse myeloma chromatin.
SDS gel electrophoresis was carried out in a
discontinuous system after nuclease digestion
of some of the chromatin DNA. From Murphy,
et al. (25).

systems of sodium dodecyl sulfate electrophoresis were
employed by the two groups and so extension of these
results to indicate clear-cut differences between either
the two tissues or the two chromatin fractionation systems is not justified. It is apparent that, in addition
to a higher content of nonhistone proteins, transcribable chromatin contains different nonhistones from those
present in repressed segments of the genome.
    The compositional studies of fractionated chroma-
tin can conclude with good chemical evidence which confirms and extends the cytochemical conclusions on the

protein contents of these two types of chromatin. The histone content of transcribable chromatin is slightly but significantly lower than that of repressed chromatin. This difference is primarily due to a decrease in the content of f1, the lysine-rich histone, in the transcribable segments. When the dimeric form of f3 histone is present in a given chromatin sample, it is confined to the condensed, repressed portion of chromatin. Transcribable chromatin is associated with an increased content of nonhistone protein and the nonhistones present in transcribable chromatin differ both qualitatively and quantitatively from those associated with repressed chromatin segments.

Structure of Chromatin Fractions

In consideration of the structure of transcribable and repressed chromatin, some complexities not present in our compositional analysis arise. Primarily, these are associated with the fact, now well documented, that the conformation of DNA (which dominates the structure of chromatin) is different for DNA in a protein-free state and DNA in the chromatin complex (for reviews see 8 and 41). Hence, it will be necessary to include a brief review of chromatin structure during the discussion of the structural properties of the two different types of chromatin we are considering today. I will compare the physicochemical properties of unfractionated chromatin, repressed chromatin segments, transcribable chromatin, and those of protein-free DNA.

I noted above that the melting properties of early and late eluted chromatin fractions from ECTHAM-cellulose chromatography differed from both unfractionated chromatin and free DNA (Figures 1,9). Similar general features of this property of repressed and transcribable chromatins have also been obtained in chromatin fractionated by other means. Thus, melting of calf thymus chromatin fractionated on sucrose gradients by Chalkley and Jensen (6) and hepatoma chromatin by Duerksen and McCarthy (11) led to increased melting temperatures for repressed fractions and decreased melting temperatures for transcribable fractions, although not to quite the degree obtained in the ion exchange fractionation (36). The stabilization to melting of a nucleic acid double helix probably reflects most directly the direct interaction of basic protein groups with the acidic phosphates of the DNA. Hence, the decrease in melting temperature for the transcribable fractions of chromatin is most likely a reflection of the diminution in content of f1 histone. Indeed, similar decreases in melting temperature have been observed for chromatin samples depleted of their content of f1 histone by extraction with 0.6 M NaCl by Ohlenbusch and coworkers (28).

Another spectroscopic property of DNA which is altered when chromatin is compared to free DNA is the circular dichroism (CD) spectrum of the nucleic acid (38,43). The CD spectrum of free DNA is characterized by a positive ellipticity band at about 275 nm with ellipticity of 9500 deg cm^2/dmole at low ionic strengths and an equal sized negative ellipticity band at about 245 nm. For DNA in the chromatin complex, there appears to be little alteration of the negative, 245 nm, band but the positive longer wavelength band is split and the maximal ellipticity is markedly reduced to a value of about 4000-5000 deg cm^2/dmole (Figure 20). The decrease in ellipticity reflects some alteration of the conformation of DNA from the normal B form characteristic of protein-free DNA to some other conformation. Some workers have tentatively assigned this altered conformation as the assumption of a C type structure for DNA in chromatin. Others have proposed supercoiling of the DNA double helix into some form of higher order structure when the nucleic acid is complexed with histones in chromatin. Whatever the cause for the altered CD spectrum of chromatin DNA in the region from 260-300 nm, it has been shown to be characteristic of unfractionated chromatin, the bulk of which is repressed.

Indeed, on examination of the CD spectra of rabbit liver chromatin fractions which elute early from ECTHAM-cellulose (34), it is found that their maximal ellipticity in this region is even less than that seen for unfractionated chromatin, that is, about 3000 deg

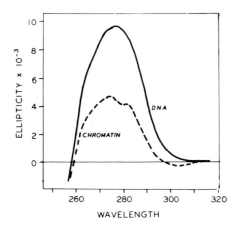

Fig. 20. Circular dichroism spectra of calf thymus
chromatin (----) and its constituent DNA
(———). Spectra were obtained in 0.001 M
Tris·Cl, pH 8.0. The region of the 245 nm
ellipticity band which is not shown was essentially identical for the two samples.
Ellipticities are expressed as molecular
ellipticities based on the concentration of
DNA-phosphate.

cm^2/dmole (Figure 21). In striking contrast, the late
eluted chromatin, thought to be more readily transcribable *in vivo*, has an ellipticity much greater than unfractionated chromatin, 6000 deg cm^2/dmole at 278 nm.
This value is about midway between the ellipticities
of repressed chromatin on the one hand and protein-free
DNA on the other. There are also slight differences in
the conformation of the proteins of the two chromatin
fractions, as evidenced by altered rotations in the
225 nm region, associated with an increased helical
content for the proteins of early eluted, repressed
chromatin (Figure 21). Similar alterations in CD spectrum are seen for calf thymus chromatin, although the
enhanced ellipticities for the late eluted fractions
are not as great as for liver chromatin. Results similar to these in terms of the DNA rotational strengths
have been obtained for calf thymus chromatin fractiona-

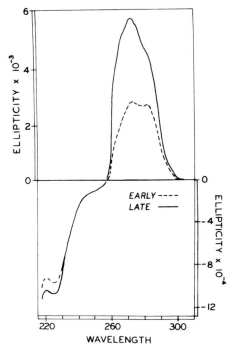

Fig. 21.   Circular dichroism spectra of early (----)
and late (———) eluted ECTHAM-cellulose fractions of rabbit liver chromatin. Spectra
were obtained in 0.25 mM EDTA, pH 7.0. Note
the difference in scales for the two portions
of the figure.

ted on sucrose gradient centrifugation by Slayter and
collaborators (44).

The implications of these findings for the structure of chromatin are unclear, largely due to uncertainty as to just what feature of the conformation of
chromatin DNA produces the altered rotational properties. Whether assumption of the C form or supercoiling
of the DNA is the cause of the altered CD spectrum of
chromatin DNA, it is apparent that the conformation of
repressed chromatin DNA is more markedly different from
that of protein-free DNA than is unfractionated chromatin. Further, the conformation of transcribable chromatin segments is altered from this unusual structure

to be more nearly like the structure of extended, free DNA. The implications of this finding for the mechanism of repression in animal cells are certainly consistent with models in which an altered DNA conformation accompanies repression, while transcribable DNA is thought to exist in a conformation more nearly similar to that of protein-free nucleic acid.

That repressed chromatin DNA exists in a different conformation from that of protein-free DNA is supported by studies of other physicochemical properties of chromatin. Since the bulk of the sequences in unfractionated chromatin are probably repressed, the conformation of whole chromatin reflects primarily the conformation of the nucleic acid in repressed segments. Early investigations of the X-ray diffraction patterns of fibers drawn from chromatin gels suggested that additional orders of structure were present in the nucleoprotein complex vs DNA alone. Thus, a series of reflections with spacings of 22, 27, 35, 55, and 105 Å were observed for chromatin but not for DNA alone (15,31,37,47). These were interpreted to indicate the presence of a supercoiled conformation for the nucleic acid of chromatin, wherein the DNA double helix underwent further coiling to form a superhelix with a pitch of 100 Å and a diameter of 120 Å. This observation is obviously consistent with many of the electron microscopic observations of chromatin fibrils.

A further striking confirmation of the altered structure of chromatin DNA vs protein-free DNA derives from studies of the flow dichroism of the two species. When oriented in a flow stream, extended molecules such as DNA exhibit dichroic absorption of light polarized parallel and perpendicular to the direction of particle orientation. As initially described by Ohba (27) and confirmed by others, the flow dichroism of chromatin DNA is of the same sign as that of isolated DNA, but only about 25% its magnitude. In part this arises from a poorer orientation of the shortened chromatin particle in the flow stream, but, in addition, the bases of DNA in chromatin appear to be less well oriented perpendicular to the long axis of the particle, consistent

with some form of supercoiling or folding of the DNA in
the nucleoprotein complex. In general, flow dichroism
and X-ray diffraction studies have correlated well with
circular dichroism investigations of the conformation
of different chromatins. Hence, with the limitations
always present in such extrapolations, it seems appropriate to suggest that the altered CD spectra of repressed and transcribable chromatin fractions derive
from changes in the degree of supercoiling or folding
of the nucleoprotein and that transcribable chromatin
is more DNA-like in conformation, while repressed chromatin possesses the supercoiled conformation which char-
acterizes unfractionated chromatin.

Hydrodynamic studies have also suggested that the
conformation of DNA in chromatin differs from that of
the isolated nucleic acid. Early ultracentrifugal and
light scattering studies of Zubay and Doty (50) demonstrated that the radius of gyration of chromatin DNA
was smaller than that of free DNA of equivalent molecular weight, consistent with some form of folding of the
nucleic acid when complexed with histones. Of the methods of separation which I have described, two depend on
a difference in the conformation of DNA in repressed
and transcribable chromatin segments. Thus, both the
sedimentation and gel filtration methods separate transcribable chromatin, in a more extended conformation,
from repressed chromatin, in a relatively condensed
structure. This supposition about the structures of
these two species has been reinforced in our preliminary studies of the hydrodynamic properties of chromatin enriched in transcribable and repressed segments
by ion exchange chromatography (I. Polacow and R.T.
Simpson, unpublished observations). Here we select by
criteria other than conformation or size and yet we
seem to find conformational properties of the two fractions which parallel those found by others whose separation method depends on size or shape.

Light scattering studies of early and late eluted
calf thymus chromatin fractions in 0.7 mM sodium phosphate, pH 6.8, indicate roughly similar molecular
weights for the two species, 2.6×10^6 and 2.0×10^6 daltons

for the two species respectively (Figure 22). The

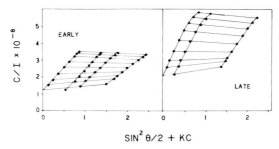

$$\text{SIN}^2 \theta/2 + \text{KC}$$

Fig. 22.   Zimm plots for light scattering by early and
late eluted ECTHAM-cellulose fractions of
calf thymus chromatin, as indicated. Measurements were made in 0.7 mM sodium phosphate,
pH 6.8.

square of the radius of gyration, R_G^2, for transcribable
chromatin is 1.5 times that for repressed chromatin, in
spite of a somewhat smaller molecular weight, indicating a more extended conformation for the DNA of transcribable chromatin. Band sedimentation velocity experiments for the extremes of the spectrum of chromatin
types obtained by ion exchange chromatography also suggest this conclusion. Thus, when sedimented in self-
generating D_2O density gradients containing 0.01 M
Tris·Cl, pH 8.0, the mean sedimentation coefficient for
repressed chromatin is 21 S while that for transcribable
chromatin is 9 S. A similar difference in sedimentation
behavior is observed for early and late eluted chromatin
fractions sedimented in sucrose gradients (Figure 23).
Early eluted chromatin sediments on the centrifugal and
late eluted chromatin on the centripetal side of the
peak observed for sonicated, unfractionated chromatin.
Since the particle weights are apparently nearly identical, the higher sedimentation coefficient for the repressed chromatin must reflect a more condensed conformation than the extended structure of transcribable
chromatin segments.
In any physical investigation, nothing is more
satisfying than the direct physical observation of the

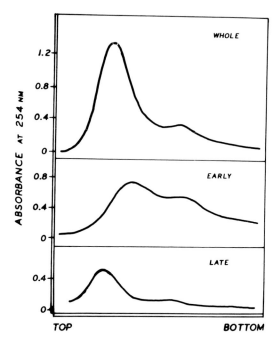

Fig. 23.   Sedimentation patterns of sonicated chromatin
and early and late eluted fractions of rabbit
liver chromatin fractionated on ECTHAM-cellulose.   Sedimentation was carried out in 5 -
50% (w/v) sucrose gradients containing 0.01 M
Tris·Cl, pH 8.0, using an SW27 rotor at
23,000 rpm for 15 hr at 2°.

conformation of a particle in the electron microscope
and therefore it is pleasing to note that the conformations predicted by solution chemical approaches have
indeed been observed by electron microscopy for two
types of fractionated chromatin.   Slayter and coworkers
(44) have studied the microscopic appearance of calf
thymus chromatin fractions obtained by sucrose gradient
sedimentation.   While the morphological characteristics
of the fractions varied widely dependent on the ionic
strength of preparation or fractionation, there was a
definite trend towards the rapidly sedimenting (repressed)
fractions being broader, more aggregated species, while

the slowly sedimenting (transcribable) fractions were
narrower, more extended fibrils (Figure 24). Michael

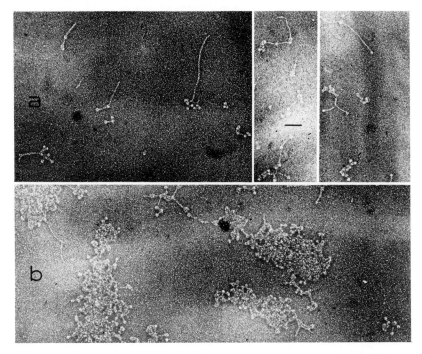

Fig. 24.   Electron micrographs of slowly (a) and
rapidly (b) sedimenting fractions of calf
thymus chromatin. Samples were centrifuged
on linear 5 - 10% sucrose gradients in 0.005
M NH$_4$OAc and rotary shadow cast with platinum.
The bar indicates 1000 Å. From Slayter, *et
al.* (44).

Beer and coworkers (unpublished observations) have
recently obtained electron micrographs of rabbit liver
chromatin fractions separated by ECTHAM-cellulose chromatography. Transcribable chromatin seems to exist pre-
dominantly as 30 Å wide fibrils, nearly as extended as
free DNA. In marked contrast, repressed chromatin is
condensed into wider fibers, with an average diameter
of about 100 Å. The fibril diameter of transcribable
chromatin is near to that of protein-free DNA while that

for repressed chromatin is near to that of the postulated supercoil. These observations will obviously be
relevant later when we consider the relation of these
fractionation studies to current models for the restriction of transcription in animal cells. At this point,
it will suffice to say that all of the physicochemical
evidence concerning the structure of the chromatin fractions is consistent with our initial hypothesis, namely
that the repressed chromatin segments are folded or
supercoiled into a highly condensed conformation, while
the transcribable segments appear to exist in a more
extended conformation, similar to that of protein-free
DNA alone.

Let me now speak briefly of another project involving chromatin fractionation that is currently underway
in our laboratory - one which is at a very preliminary
stage but which holds some interest in terms of further
understanding the differences between repressed and
transcribed chromatin. We are investigating the mode
of histone binding to DNA in repressed and transcribable
chromatin fractions. In view of the differences in both
functional and structural properties between these two
types of chromatin, the seeming constancy of all the
histones save f1 was unexpected. Perhaps the same histones were there but their interaction with DNA differed in the two species. Our route of investigation
of this question is one which we have already employed
in studying histone binding to DNA in unfractionated
chromatin (39). Isolated chromatin is chemically modified with acetic anhydride - this agent will acetylate
lysyl residues, but only when they are not bound to DNA.
Hence the degree of lysine modification is a quantitative measure of the number of lysyl residues not bound
to the nucleic acid. Isolation of the histones followed by sequence studies should allow the determination
of just what regions of given histones are and are not
bound to DNA in the native chromatin complex.

Based on these and other studies (40), we proposed
previously that histone binding to DNA in chromatin
might occur at both ends of the histone molecule with
the middle region of the protein looped out from the

nucleic acid (Figure 25). Similar conclusions concerning the microscopic details of the binding of histones
to DNA have been reached by Bradbury and coworkers on
the basis of nuclear magnetic resonance studies of histone–DNA complexes (3,4) and by Richards and Pardon
(37) from consideration of the primary sequence of one
of the histones. We have further postulated that this
type of bridging or inchworm binding might constrain
the DNA by forming crosslinks to create supercoiled or
folded structures (40).

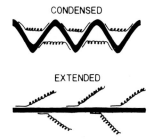

CONDENSED

EXTENDED

Fig. 25. Model for histone binding to condensed and
extended chromatin. Condensed chromatin is
represented as a supercoil stabilized by
bridging histone molecules bound at both
their amino and carboxyl termini but non-
bound in the central region. Extended chromatin is shown with histones bound only by
their amino terminal regions.

If transcribable chromatin is extended, then this
might arise by the binding of only a portion of the
histone, for example, only the amino terminal end. The
bridge is lost, constraint is released, and the supercoil would then extend (Figure 25). If this were the
case, then more histone lysyl residues ought to be available for acetylation in transcribed chromatin. Our
experiments suggest that this does indeed occur. Rabbit
liver chromatin from early and late eluted fractions was
acetylated with an excess of acetic anhydride and the
histones isolated and fractionated on 25 cm long sodium
dodecyl sulfate polyacrylamide gels. After staining

and scanning, the gels were sliced and incorporation of labeled acetate determined (Figure 26). The degree of acetylation of histones f2a1 and f2a2 appears similar in both fractions. In contrast, there is markedly more

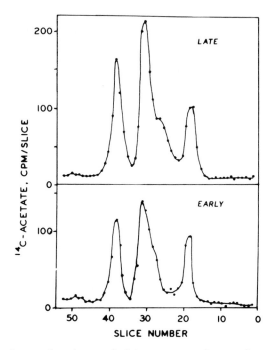

Fig. 26. Acetylation of histones of condensed and extended chromatin. Early and late eluted ECTHAM-cellulose fractions of rabbit liver chromatin were acetylated as described (39), histones were isolated by acid extraction and fractionated by electrophoresis on 15 cm SDS polyacrylamide gels. After staining and scanning, the gels were sliced and incorporated radioactive acetate determined by scintillation counting. The gels contained identical amounts of histones. Only the lower portion of the gel (excluding histone f1) is shown in the figure.

acetate associated with fractions f3 and f2b in the late eluted chromatin, suggesting that indeed these two histone fractions are bound differently in repressed and transcribed chromatin and, further, that the binding to transcribable chromatin is such that fewer of the lysyl residues are associated with DNA. Unequivocal proof of this hypothesis is a long job requiring peptide isolation and sequence determination but the initial results are quite reasonable and provocative and suggest a logical explanation of how transcribable and repressed chromatin segments may contain the same content of four of the histone classes and yet differ so strikingly in both functional and structural properties.

Let me summarize the data I have presented today. Methods based on both the conformational and charge characteristics of chromatin fractions seem to allow the separation of segments with similar characteristics. Functional and structural evidence suggests that chromatin segments which are condensed or supercoiled are repressed *in vivo* and *in vitro* while those which are in an extended, more DNA-like conformation are both transcribed and may contain the regulatory sites for genes which are active in a given tissue. Extended sequences are associated with decreased content of fl histone and the binding of two other histones may be different in extended as compared to condensed chromatin segments. Transcribable chromatin is associated with an increased content of nonhistone proteins and the nonhistone proteins present in repressed and transcribable chromatin appear to differ both qualitatively and quantitatively.

Models for Eukaryotic Gene Regulation

Let us now attempt to put this all together within the context of two models for the mechanism of gene regulation in animal cells. Of the currently available hypotheses for gene regulation in eukaryotes, only Crick and Paul have suggested concrete structural features for the nucleoprotein. Crick (9) has postulated a model in which transcribable chromatin exists as extended fibrous strands and repressed chromatin exists

177

as contiguous globular regions, containing the control
DNA sequences in single stranded array (Figure 27).

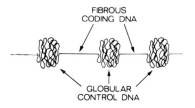

Fig. 27.   The Crick model for chromosome structure.
From Crick (9).

The data are inconsistent with this hypothesis on several grounds. First, nascent mRNA appears to be associated with both extended and condensed chromatin, albeit more so in the former. Secondly, control regions,
at least those which bind hormone receptor proteins,
are associated with transcribable chromatin and not
with condensed segments. Finally, in an immunochemical
study of the amount of single stranded DNA in chromatin
we have been able to show that the amount of single
stranded nucleic acid present is far less than that con-
servatively estimated as necessary for control loci (20).
     The model of John Paul (32), on the other hand,
provides a far more convincing agreement with ours and
others' data. The Paul model postulates that the control regions of chromatin are in an extended array and
that the remainder of the nucleic acid is supercoiled
(Figure 28). The control sites, including repressor
binding sites, polymerase binding sites and specific
control loci are linked to their structural genes.
Specific anionic molecules, probably nonhistone proteins,
are thought to bind to a region on the DNA and create
the extended configuration of the control region. As
polymerase transcribes a structural gene, transient re-
laxation of supercoiling is thought to occur. The model
is consistent with the data in the following ways. Con-
trol-site binding proteins, *i.e.*, hormone receptor proteins, bind to the extended segments of chromatin.
Transcription *in vivo* is observed, in our hands, on both

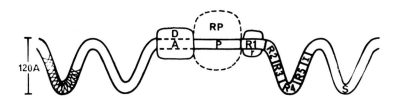

Fig. 28.   The Paul model for chromosome structure.
A = address site. D = destabilizer molecule,
possibly a nonhistone protein. P = promoter
site. RP = RNA polymerase. R1 - R5 = regulator sites. r = repressor. I = initiation
site. S = structural gene. From Paul (32).

extended and condensed chromatin. The amount of synthesis increases as one goes more and more to the extended
segments, consistent with initiation at these sites and
the presence of terminators such that all segments of
supercoiled chromatin are not transcribed. Extended
regions contain increased amounts of specific nonhistone
proteins, possibly the anionic molecules postulated to
open up the conformation of the DNA. About the only
additions to the model required to produce complete consistency with the data are the presence in extended
chromatin of altered histone binding, possibly due to
the nonhistone proteins present, and the inclusion of
decreased fl histone content in the extended sequences.
This latter inclusion is of particular relevance. In
general, the histones of a given cell are quite inactive
metabolically. Once synthesized, they appear to remain
associated with DNA and have no demonstrable metabolic
turnover (see 17). fl histone, in contrast, does turn
over (16). This finding would be that expected if, during the turning on and off of genes, the bulk of the
histones were not dissociated from DNA but merely altered in their binding to the nucleic acid while fl histone, during the same process, was dissociated and later
replaced, possible after turnover.

Let me conclude by quoting from an Editorial
which appeared in *Nature* about a year and a half ago.

"Just as the structure of DNA was the basic problem which had to be solved before molecular biology became a meaningful discipline, so the structure of the eukaryotic chromosome is the vital issue which must be resolved before research today in cell biology can produce a coherent set of concepts instead of a mass of unrelated data." I agree strongly with the sentiment and hope that the investigations which I have presented today are bringing us closer to the achievement of that goal.

Acknowledgment

The studies from our laboratory are the product of interactions between Drs. Gerald Reeck, Ione Polacow, Ronald Seale, Herbert Sober, and myself. Mrs. Bessie Watkins and Ms. Linda Propst have provided valuable technical assistance during our studies. I am grateful to Dr. Edwin Murphy for communication of his results prior to publication.

References

1. Ansevin, A.T., Hnilca, L.S., Spelsberg, T.C. and Kehm, S.L. Structure studies on chromatin and nucleohistones. Thermal denaturation profiles recorded in the presence of urea. Biochemistry 10, 4793 (1971).

2. Bonner, J., Dahmus, M.E., Fambrough, D., Huang, R.C., Marushige, K.,and Tuan, D.Y.H. The biology of isolated chromatin. Science 159, 57 (1968).

3. Boublik, M., Bradbury, E.M., Crane-Robinson, C. and Rattle, H.W.E. Proton magnetic resonance studies of the interaction of histones F1 and F2B with DNA. Nature New Biol. 229, 149 (1971).

4. Bradbury, E.M., Carpenter, B.G. and Rattle, H.W.E. Magnetic resonance studies on deoxyribonucleoprotein. Nature 241, 123 (1973).

5. Britten, R.J. and Kohne, D.E. Repeated sequences in DNA. Science 161, 529 (1968).

6. Chalkley, R. and Jensen, R.H. A study of the structure of isolated chromatin. Biochemistry 7, 4380 (1968).

7. Clark, R.J. and Felsenfeld, G. Structure of chromatin. Nature New Biol. 229, 101 (1971).

8. Comings, D.E. The structure and function of chromatin. Adv. Human Genetics 1972, 237 (1972).

9. Crick, F. General model for the chromosomes of higher organisms. Nature 234, 25 (1971).

10. DeLange, R.J. and Smith, E.L. Histones: structure and function. Ann. Rev. Biochem. 40, 279 (1971).

11. Duerksen, J.D. and McCarthy, B.J. Distribution of deoxyribonucleic acid sequences in fractionated chromatin. Biochemistry 10, 1471 (1971).

12. Elgin, S.C.R. and Bonner, J. Limited heterogeneity of the major nonhistone chromosomal proteins. Biochemistry 9, 4440 (1970).

13. Frenster, J.H. Ultrastructural continuity between active and repressed chromatin. Nature 205, 1341 (1965).

14. Frenster, J.H., Allfrey, V.G. and Mirsky, A.E. Repressed and active chromatin isolated from inter-phase lymphocytes. Proc. Nat. Acad. Sci. U.S. 50, 1026 (1963).

15. Garrett, R.A. A physical study of the stability of the native nucleohistone conformation to salt dissociation and heating. Biochemistry 10, 2227 (1971).

16. Gurley, L.R. and Hardin, J.M. The metabolism of histone fractions. III. Synthesis and turnover of histone f1. Arch. Biochem. Biophys. 136, 392 (1970).

17. Hnilica, L.S. The Structure and Biological Functions of Histones. CRC Press, Cleveland, Ohio, 1972.

18. Janowski, M., Nasser, D.S. and McCarthy, B.J. Fractionation of mammalian chromatin. Acta Endocrinol. 1972, 112 (1972).

19. Johnson, R.S., Chan, A. and Hanlon, S. Mixed conformations of deoxyribonucleic acid in intact chromatin isolated by various preparative methods. Biochemistry 11, 4347 (1972).

20. Levy, S. and Simpson, R.T. Quantitative immunochemical search for single stranded DNA in chromatin. Nature New Biol. 241, 139 (1973).

21. Li, H.J. and Bonner, J. Interaction of histone half-molecules with deoxyribonucleic acid. Biochemistry 10, 1461 (1971).

22. Littau, V.C., Allfrey, V.G., Frenster, J.H. and Mirsky, A.E. Active and inactive regions of nuclear chromatin as revealed by electron microscope autoradiography. Proc. Nat. Acad. Sci. U.S. 52, 93 (1964).

23. Matsuyama, A., Tagashira, Y. and Nagata, C. A circular dichroism study of the conformation of DNA in rat liver chromatin. Biochim. Biophys. Acta 240, 184 (1971).

24. McConaughy, B.L. and McCarthy, B.J. Fractionation of chromatin by thermal chromatography. Biochemistry 11, 998 (1972).

25. Murphy, E.C., Jr., Hall, S.H., Sheperd, J.H. and
     Weiser, R.S. Fractionation of mouse myeloma
     chromatin. Biochemistry 12, 3843 (1973).

26. Nishiura, J.T. Ph.D. Dissertation, Department of
     Genetics, University of Washington, Seattle, Wash-
     ington, 1972.

27. Ohba, Y. Structure of nucleohistone. I. Hydro-
     dynamic behavior. Biochim. Biophys. Acta 123, 76
     (1966).

28. Ohlenbusch, H.H., Olivera, B.M., Tuan, D.Y.H.
     and Davidson, N. Selective dissociation of his-
     tones from calf thymus nucleoprotein. J. Mol.
     Biol. 25, 299 (1967).

29. O'Malley, B.W., Toft, D.O. and Sherman, M.R.
     Progesterone-binding components of chick oviduct.
     II. Nuclear components. J. Biol. Chem. 246, 1117
     (1971).

30. Panyim, S. and Chalkley, R. The heterogeneity
     of histones. I. A quantitative analysis of calf
     histones in very long polyacrylamide gels. Bio-
     chemistry 8, 3972 (1969).

31. Pardon, J.F., Wilkins, M.H.F. and Richards, B.M.
     Superhelical model for nucleohistone. Nature 215,
     508 (1967).

32. Paul, J. General theory of chromosome structure
     and gene activation in eukaryotes. Nature 238,
     444 (1972).

33. Peterson, E.A. and Kuff, E.L. Chromatographic
     isolation of 80 S ribosomes from rat liver and
     mouse plasma cell tumor. Biochemistry 8, 2916
     (1969).

34. Polacow, I. and Simpson, R.T. Circular dichroism spectra of putative transcribed and repressed chromatin. Biochem. Biophys. Res. Comm. 52, 202 (1973).

35. Reeck, G.R. and Simpson, R.T. ECTHAM-cellulose fractionation of chromatin: functional properties of the isolated species. In preparation.

36. Reeck, G.R., Simpson, R.T. and Sober, H.A. Resolution of a spectrum of nucleoprotein species in sonicated chromatin. Proc. Nat. Acad. Sci. U.S. 69, 2317 (1972).

37. Richards, B.M. and Pardon, J.F. The molecular structure of nucleohistone (DNH). Exptl. Cell Res. 62, 184 (1970).

38. Shih, T.Y. and Fasman, G.D. Conformation of deoxyribonucleic acid in chromatin. J. Mol. Biol. 52, 125 (1970).

39. Simpson, R.T. Modification of chromatin with acetic anhydride. Biochemistry 10, 4466 (1971).

40. Simpson, R.T. Modification of chromatin by trypsin. The role of proteins in maintainance of deoxyribonucleic acid conformation. Biochemistry 11, 2003 (1972).

41. Simpson, R.T. Structure and function of chromatin. Adv. Enzymol.,38, 41 (1973).

42. Simpson, R.T. and Reeck, G.R. A comparison of the proteins of condensed and estended chromatin fractions from rabbit liver and calf thymus. Biochemistry 12, 3853 (1973).

43. Simpson, R.T. and Sober, H.A. Circular dichroism of calf liver nucleohistone. Biochemistry 9, 3103 (1970).

44. Slayter, H.S., Shih, T.Y., Adler, A.J. and Fasman, G.D. Electron microscopic and circular dichroism studies on chromatin. Biochemistry 11, 3044 (1972).

45. Spelsberg, T.C., Steggles, A.W. and O'Malley, B.W. Progesterone-binding components of chick oviduct. III. Chromatin acceptor sites. J. Biol. Chem. 246, 4188 (1971).

46. Steggles, A.W., Spelsberg, T.C., Glasser, S.R. and O'Malley, B.W. Soluble complexes between steroid hormones and target-tissue receptors bound specifically to target-tissue chromatin. Proc. Nat. Acad. Sci. U.S. 68, 1479 (1971).

47. Wilkins, M.H.F., Zubay, G. and Wilson, H.R. X-ray diffraction studies of the molecular structure of nucleohistone and chromosomes. J. Mol. Biol. 1, 179 (1959).

48. Yunis, J.J. and Yasmineh, W.G. Heterochromatin, satellite DNA, and cell function. Science 174, 1200 (1971).

49. Zirkin, B.R. The protein composition of nuclei during spermiogenesis in the leopard frog, Rana pipiens. Chromosoma 31, 231 (1970).

50. Zubay, G. and Doty, P. The isolation and properties of deoxyribonucleo-protein particles containing single nucleic acid molecules. J. Mol. Biol. 1, 1 (1959).

GENE EXPRESSION IN ANIMAL CELLS

E. Brad Thompson

Laboratory of Biochemistry
National Cancer Institute
National Institutes of Health
Bethesda, Maryland   20014

How should we define gene expression? Expression
of genes at what level? For this discussion gene expression shall refer to phenotype, that is whatever
level one can measure in a specific way. Control of
gene expression then includes more than control of
which genes are transcribed, but also whether the products of transcription can be expressed in a measurable
way. This is necessary, since it appears that functional controls may exist at several levels in eukaryotes. The subdivision of animal cells into various
functioning organelles makes life complicated for biol-
ogists who wish to study what genes do. The central
problem in understanding eukaryotic gene expression includes not only understanding, in molecular terms, the
selective transcription of individual genes within the
nucleus, but also how the transcription products reach
the cytoplasm and are controlled there. What this lecture will try to do is to outline the cellular organization and molecular events leading from DNA to final
molecular phenotype and to look for known or potential
sites of regulation. Later, I will try to cite some
experiments illustrating some of these levels of control.
        To illustrate its organization, Fig. 1 shows a
crude drawing of a cell with some of its parts, both
organelles and certain important macromolecules. Now
the question is, can what is known of molecular events

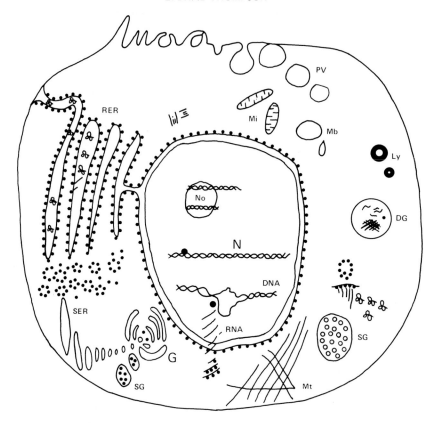

Fig. 1. Drawing of generalized mammalian cell, indicating some functional organelles and macromolecules. Abbreviations: PV, pinocytotic
vesicle; RER, rough endoplasmic reticulum;
Mi, mitochondria; Mb, microbody; Ly, lysosome;
DG, digestive granule; SG, secretory granule;
Mt, microtubule; G, Golgi apparatus; SER,
smooth endoplasmic reticulum; No, nucleolus;
N, nucleus; DNA, deoxyribonucleic acid; RNA,
ribonucleic acid. (Reproduced with permission
from ref. 41)

explain regulation at all the levels the cell displays?
Consider some of these. 1) *Molecular level:* Naturally
as biochemists we imagine all levels ultimately to be
explicable in terms of molecules. But by mentioning

this specific category I mean to call attention to the
variety of molecular phenotypes. It is thought that
all the cells of a given animal contain the same or
nearly the same complement of genes. Yet we all know
that different cells contain different proteins and
RNAs. Some are seen in all cells; others only in certain classes of cells. Of the proteins which are present in a given cell, some are constitutive and others
are variable in amount. In short, what are the mechanisms for differentiated molecular behavior?

Chromosomal level: The DNA of animal cells is
associated with proteins in specific units, the chromosomes. They are readily identified as discrete objects
at metaphase, when transcription is minimal, but grow
morphologically more vague during interphase, when most
transcription occurs. The relation between DNA strands
in interphase and metaphase chromosomes has been shown
in a variety of experiments. But the exact steps by
which the correct interphase DNA strands are condensed
into their chromsomes are not known. All of mammalian
genetic mapping is referred to the chromosomal units
and they behave genetically as linear arrays of genes.
Yet the exact arrangement of the DNA in each chromosome
is not clear, not even as to whether there are one or
several strands. Although there appears to be in chromatin a single fiber of about 30 Å in diameter consisting of DNA double helix along with associated proteins,
there is some controversy whether there is one such
fiber per chromosome or more than one. In addition to
that fiber there seems to be an additional level of
structuring; it is as though the single fiber in many
areas is looped again into what is called supercoiling.
The diameter of these coils is about 200 Å, and they are
spaced something like 100 Å apart. Whether supercoiling
merely represents a way to place conveniently a lot of
DNA into a very small space or whether it has implications for control is not clear yet.

Furthermore, chromosomes vary in activity. Some
are almost entirely silent, carrying only a few genes
which are expressed. The Y chromosome in the male is
an example. So is one of the two X chromosomes of the

female (26). Somehow, the cell has a mechanism for randomly and permanently inactivating all but one of its X-chromosomes. Then there are position effects. Sometimes a piece will break off one chromosome and attach itself to another. Whether the huge number of genes on this reattached piece express themselves or not depends on the state of the chromosome to which it attaches. If the recipient chromosome is largely inactive, it tends to depress or block entirely the gene expression of the attached fragment. If conditions are right, a gradient of effect down the fragment can be demonstrated. The extent of the entire position effect may depend as well on the state of general activity and dose of other chromosomes in that cell (19).

Cellular organelle level: For example, when the pyknotic, inactive nucleus of an avian erthrocyte is made to enter a vigorously growing HeLa cell, the erythrocyte nucleus gradually swells and again becomes active in RNA synthesis (18).

Cellular level: The growth cycle and the morphology of various cultured cells, controlled by small molecules such as cyclic AMP and hormones, are examples of this phenotypic level of control (33).

Multicellular level: Obviously cells form tissues, associating with one another in a regulated manner. In tissue culture it can be demonstrated that disaggregated mixtures of certain cells will sort themselves automatically into cell types.

It is by no means clear how the known molecular events explain all these phenomena. But perhaps by considering the molecular organization of chromatin we can discuss some principles of control. Chromatin consists of DNA, histones, non-histone proteins and RNA in the rough proportions of 1 : 1 : 0.5 : 0.05. The close ratio between DNA and histones occurs with little variation from cell to cell. Furthermore the ability of the cell to make DNA and histones is linked very closely with histones synthesized in the cytoplasm and transported to the nucleus during S phase, the period of DNA synthesis. Non-histones, that is whatever proteins are left over after extracting histones, occur in

a broad range of proportions in different tissues.
Many are tissue-specific. Then there is a little RNA
which also is variable from tissue to tissue in a dis-
trinctive way.

The DNA in this material is known to contain several sorts of regions. First, there are structural
genes for proteins. These appear to be represented
only once or a few times in each haploid set of DNA
(32,40). A rare exception are the genes for histones.
These are repeated several hundred, perhaps as many as
1000 times. Second, there are the genes coding for
ribosomal and transfer RNA. These genes are redundant
and occupy a few percent of the total DNA. Ribosomal
genes are clustered in the nucleolar organizer while
tRNA genes appear to be disperse. Third are more highly repetitious regions which consist of families of
short stretches of DNA very similar to one another but
not identical. These regions of short repetitious DNA
are scattered throughout the DNA. One particularly
highly repetitious class occurs as a group and is associated with the constitutive heterochromatin of the
centromere. The size and base composition of repeti-
tive DNA indicate that it is probably not coding for
meaningful peptides and much of it is probably never
translated (27).

The histones are a family of basic proteins rich
in arginine and lysine which bind tightly to DNA. They
have been extensively characterized in many tissues
and the primary structure of some has been obtained.
There are 6 or 7 specific histone fractions, separable
on the basis of their relative lysine/arginine content.
Each fraction seems to be a unique species, except the
lysine-rich F1 histone. This appears to have some microheterogeneity, probably on the basis of modified side
groups. Histones are found to be modified by acetylation, phosphoylation, methylation, and reversible oxidation of thiol groups. Several histones have been
shown to possess a clustering of their basic amino
acids at one end of the molecule. Since it is by way
of these basic groups that they react with the phosphate groups of DNA, this means that one end of the

molecule may bind while the other is relatively free.
At first it was thought that histones might be the regulatory proteins accounting for transcription control
of specific unique genes. For several years the literature has been full of papers studying numerous regulated molecular events, stating that such and such
increase in enzyme activity was probably due to release
of repression by removing a histone from a specific
gene. Now the consensus appears to be that they are
probably not the fine regulatory elements. Objections
to histones as specific gene regulators include the
following: They are too few in number. They occur
in similar proportions in all tissues. They are found
in equal amounts in regions of chromatin actively engaged in RNA synthesis (euchromatin, Balbiani rings)
as in quiescent chromatin (heterochromatin)(27). There
are not good correlations between variations in histones in differentiated cells.

At present, three possible functions are popularly
ascribed to histones. They may assist with the structural arrangement of DNA or in regulating its replication. They might also be involved in preventing transcription of large segments of DNA, as in the constitutive heterochromatin of certain chromosomes. Lastly,
one physical model has suggested that they participate
in the super-coiling of DNA, which configuration blocks
the access of RNA polymerase (21). It has been suggested that modification of histones, that is reversible thiol oxidation, phosphorylation and so forth
might provide the necessary variety and specificity
which seem lacking, and many correlations between various biologic effects and these events have been noted,
but the correlations have not as yet become compelling.
Since one of the other lectures in this book deals with
the problem of nucleoproteins and control in more detail, I will not take further time to discuss histones
here.

The non-histone acidic proteins of chromatin (or
of the nucleus) are those proteins left over after one
removes histones, for instance by acid extraction.
They are a very heterogeneous group, and include pro-

teins transiently entering the nucleus from the cytoplasm (functionally or as preparative artefacts), various nuclear enzymes, and proteins which associate closely with DNA. Part of the nonhistone nuclear proteins are species and organ specific. They have not been studied as thoroughly as histones, but are being considered as strong candidates for regulatory function because of their variety, more rapid turnover, and specificity. Also the possible analogy with the two well-studied prokaryote repressors, lac and λ, has been frequently mentioned (27).

A special class of the non-histone proteins needs mentioning. This constitutes the RNA polymerases of animal cells. For comparison, we may recall that *E. coli* RNA polymerase has been shown to be a large enzyme, M.W. about 5×10^5, and consisting of five subunits, $\alpha_2\beta\beta'\sigma$. The M.W. of each α subunit is 40,000, while β and β' are 155,000 and 165,000 respectively. The initiation-site specific subunit is σ, M.W. about 90,000. Transcription consists of the four steps, binding of enzyme to DNA, initiation of RNA synthesis, elongation of the RNA polymer, and termination, which requires the additional protein factor ρ. Although the core enzyme, that is, $\alpha_2\beta\beta'$, can bind and transcribe, it does so poorly and inaccurately. Transient association with σ allows specificity and efficiency. At first it was thought that there might be many σ's, each providing specific initiation at certain operons, but in fact there appear to be only a few, and fine regulation must be provided by other means.

In animal cells, less is known. However, it is clear that there is more than one DNA polymerase. In the nucleus as many as four have been reported (35). Clearly, there are distinctly different enzymes in nucleus and nucleolus. Like their prokaryotic counterparts, these also are large, multiple-subunit enzymes (9). They differ in their activation by Mn^{++} and Mg^{++}, their response to various templates, and their inhibition by α-amanitin (20). In addition, mitochondria have their own RNA polymerase, a small enzyme with M.W. about 64,000 (23).

The remaining macromolecular component of chromatin is RNA. The existence of this RNA as a special functional class of RNA, rather than adventitious material such as tRNA, is controversial. The proposal has been made, however, that it binds covalently to histones and/or acidic proteins and hydrogen bonds to double-stranded DNA, thereby providing specificity for recognition of regulatory sequences in the DNA (14).

The areas of contrast between the organization of prokaryotic and eukaryotic genetic material are obvious. Eukaryotes differ in possessing DNA which is organelle-restricted, which contains many redundant sequences, and which is closely associated with a large number of basic and acidic proteins. Eukaryotic DNA differs also in that most of it in a given cell is never transcribed but is permanently repressed. No doubt the huge quantity of DNA in eukaryotes requires special proteins, to provide the structure necessary for packing.

When we consider transcription, further contrasts appear. In prokaryotes, as a cistron is transcribed and a mRNA chain develops, ribosomes immediately associate with it and begin synthesizing protein. Transcription and translation of a strand of mRNA proceed simultaneously. In eukaryotes the processes are physically and temporally separated. Actually, most of the very rapidly-labelled RNA in animal cells is synthesized and degraded without ever leaving the nucleus (17), a fact that was relatively neglected for several years. The RNA of rapid turnover is large, up to 70S or more, and has a broad range of sizes. Therefore, it is sometimes referred to as heterogeneous nuclear RNA (HnRNA)(14a). Such molecules are far too large to represent the information for a single protein of normal size. Earlier kinetic studies showed that most HnRNA never left the nucleus (4), but more recently it has begun to seem that it includes the mRNA fraction. Several types of evidence support this. First, the ability of HnRNA to hybridize with DNA can be partially competed for by mRNA. This type of study and other annealing studies are consistent with the view that HnRNA includes mRNA but represents a broader range of

DNA sequences than does mRNA. Also DNA:RNA annealing competition studies show that the HnRNA differs among various tissues. It should be remembered, though, that the techniques of annealing used in these studies are frequently measuring the repeated RNA sequences, and not the unique ones in mRNA. However, some specific mRNA's have been used in such studies (29,47). A second line of evidence is found in the fact that long stretches of polyadenylic acid are found attached to the 3' end of both HnRNA and mRNA, but not to rRNA or tRNA (2). These sequences are added enzymatically, after transcription. Also, newer kinetic studies and experiments with the drug cordycepin support the suggested precursor-product relationship.

It is not known how mRNA is hidden in the pool of HnRNA. Messenger could arise from every molecule of HnRNA, or from a small subset of the general pool. In either case, the bulk of the HnRNA still would be degraded. It is also possible that mRNA which does not require trimming down from larger molecules could be hidden in the HnRNA pool, and that the large molecules of HnRNA have an entirely different and unknown function.

At any rate, the existence of this active group of large RNA molecules, among which mRNA must pass in one way or another, suggests a possible new level of control. I would propose that the phenotypic expression of a certain gene as a protein product may well be controlled by selecting in one way or another which mRNA reaches the cytoplasm. Many mRNA's, whether embodied in HnRNA or not, could be transcribed and degraded without ever leaving the nucleus. Control as to whether they obtained access to the cytoplasm could occur in several ways. Special nucleases could exist for certain classes of mRNA's or mRNA precursors. Agents which specifically enhanced or inhibited such nucleases at any level could then effect an increase or decrease in the cytoplasmic level of functional RNA. Alternatively, there could be only non-specific general nuclease activity, and protection of mRNA or its precursor in the nucleus against nucleolytic attack could come from association with special nuclear proteins and/or from modifying certain bases in the mRNA precursor. Association with such proteins and base modifications might assure survival of the mRNA till it reached the cytoplasm. Any molecule (hormone, metal ion, cAMP, etc.) which varied the synthesis or structure of the protector protein would then appear to act as a transcriptional inducer. Of course, the converse could well occur too, and instead of protector molecules there could be similar proteins which bound the nuclear mRNA in such a way that it was more readily available for degradation. Finally, there could be selection of mRNA's to enter the cytoplasm or be retained and destroyed in the nucleus at the nuclear membrane.

Several models for control at the purely transcriptional level have been proposed on the basis of the existing facts (10,13,14,16,17,49). The above suggestion in no way denies or precludes transcriptional control. The control-of-processing model is independent of other levels of control, but clearly could interact appropriately with them.

Let us now consider the steps involved between transcription and final phenotypic expression of a gene coding for some specific protein. Actually, of course, many genes are involved, because tRNA's and ribosomes must be supplied for the process. Figure 2 is a rough outline of these processes. It shows 3 genes, representing those for ribosomal RNA, transfer RNA, and a messenger RNA. Parenthetically, it should be noted that a potential manner of regulation in animal cells is gene duplication. Although some genes are multiple, as noted above, no instances of regulation utilizing gene duplication have been proven in mammals. There is, however, a beautifully studied example in the amphibian oocyte. Brown and his colleagues have shown that ribosomal cistrons are replicated many fold in this cell, enough to provide ribosomal RNA for several subsequent cell divisions (11). Therefore, in considering gene expression in mammals, one should not forget the possibility of control at the level of DNA replication.

The next level of potential control is of course the transcriptional, involving the elements discussed above and shown diagrammatically in Fig. 2. This process results in the primary gene product. It is known that with ribosomal RNA, the primary gene product is much larger than the functional rRNA in the cytoplasm. Processing occurs in the nucleus in which the rRNA precursor is stepwise degraded into smaller fragments. Thus in HeLa cells the 45S rRNA precursor is ultimately reduced in size to 18S rRNA, which reaches the cytoplasm in about 20 min., and 28S rRNA, which requires about an hour for processing. These pieces, plus the 7S and 5S fragments eventually combine to form the smaller and larger ribosomal subunits (3). Also it should be recalled that during all this processing these RNA's are in association with specific proteins and exist as ribonucleoprotein particles. Although the scheme is less well worked out, there appears to be processing involving a reduction in size of the primary gene product for tRNA's as well, as is indicated in the figure (39). We have already discussed the possibility of such being the case for mRNA's. In addition, base modifications are essential in controlling rRNA and tRNA processing and function, with specific enzymes required. Thus within the nucleus exist numerous potential regulatory sites subsequent to the actual transcription into primary gene product.

In the cytoplasm of the cell the following steps of translation take place: The mRNA with the aid of certain protein factors and trinucleotides combines with the 40S ribosomal subunit. This recognition point between message and the smaller ribosomal subunit has been suggested as a site of regulation for genes and animal cells. There may be specific recognition of certain messages in either a positive or negative sense at this level. Next, amino acids are added to the transfer RNA by tRNA-specific synthetases, and with additional protein factors the process of translation proceeds. Regulation has been suggested at the level of tRNA also. Although no clear-cut examples exist of fine scale regulation by tRNA, it is notable that in cells which are differentiated for producing certain special proteins, e.g. erythrocytes, the pattern of tRNA's has

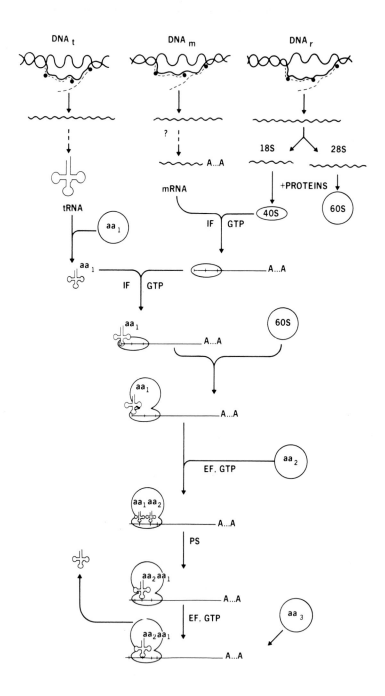

Fig. 2.   Diagram of sequential events in transcription
and translation. Transcription of three cistrons (genes) shown, one each for tRNA, mRNA
and rRNA. Subsequently, the interaction of
these elements in mRNA translation, along
with cytoplasmic factors, is shown. DNA_t,
DNA_m, DNA_r represent cistrons coding for tRNA,
mRNA and rRNA, respectively. ●---, RNA polymerase and nascent RNA chain during transcription. aa_1, aa_2, aa_3, individual amino acids.
A....A, poly adenylate sequences. IF, initiation factors. GTP, guanosine triphosphate.
EF, elongation factors. PS, peptidyl synthetase. tRNA, transfer RNA. mRNA, messenger
RNA. 18S and 28S, RNA's of the small and
large ribosomal subunits, respectively. 40S
and 60S, small and large ribosomal subunits
per se. (Reproduced with permission from
ref. 41.)

been modulated so that it better reflects the requirement of certain tRNA's for translation of specific
codons in the message for hemoglobin. Of course, since
the genetic code and tRNA's are redundant, that is,
since there is more than one tRNA and codon for each
amino acid, the possibility exists that a codon may occur in a message which is recognized only by a minor
group of the total tRNA pool. Therefore, by regulating
the level of that specific RNA one could regulate the
rate of peptide chain synthesis at this limiting point
along the message.
     When the completed protein chain is coming off the
final codon, termination must occur in response to
other specific protein factors. Recently a paper by
Oliver and his colleagues pointed out that cyclic AMP
added to a liver cell system was capable of releasing
tyrosine aminotransferase chains which apparently had
accumulated near the termination point but had failed
to be released from the polysomes (15). So termination
is a point of regulation which seems to occur at least
in this one instance. As the protein is completed,

another level of control occurs, one well-documented
in animal cells; that is the functional expression of
the completed protein, whether it be an enzyme or structural protein. Proper protein folding probably occurs
as the peptide chain nears completion or is released
(1). However, in animal cells, unlike growing bacteria,
proteins turn over; they have finite decay rates which
may vary in a single cell from minutes to days. The
regulation of this turnover is ill understood. Most
proteolytic enzymes of the cells, located almost exclusively in lysosomes, are quite general and non-
specific; therefore regulation must occur at the level
of access. That is, protein folding, ligand interactions, genetic alterations which alter lysosomes,
hormonal alterations in proteins or lysosomes, etc. are
thought to result in the varying specific degradation
rates of various proteins (37). One classic example
of control of phenotype at the level of protein decay
is given by tryptophan oxygenase, an enzyme typical of
the differentiated proteins of liver induced to increase
in amount by steroid hormones. It was observed that
tryptophan oxygenase could be induced by tryptophan
alone because the substrate stabilizes the enzyme. No
alteration in gene transcription or quantity of message
occurs, but since the enzyme is not degraded at its
usual rate, it accumulates, and the result is an altered
phenotype.

In very short-hand form, those are the elements
of transcription and translation in animal cells. Next
I'd like to discuss gene expression from a more specific
point of view. To use the inductive approach in think-
ing about the problem of gene control, I shall consider
the induction of certain products and events by steroid
hormones. I choose this field as the example for two
reasons: because it demonstrates regulation at several
of the levels discussed, and because I am familiar with
it.

Steroid hormones, administered to their appropriate target tissue, provoke altered synthesis of RNA,
DNA, and protein. For example, corticosteroids given to
an adrenalectomized rat result in enhanced incorporation of precursors into hepatic ribosomal, transfer,

and rapidly labelled cytoplasmic RNA's. There is some
stimulation of general protein synthesis, but certain
specific enzymes are induced much more than this background. On the other hand, the same steroids result
in inhibition of these processes in other cells, thymocytes and fibroblasts for instance (45). Sex steroids
cause similar trophic effects in their target tissues
(28). In general, the ability of steroids to exert
their effects is blocked by agents which inhibit RNA
or protein synthesis.

As an example of a typical steroid-induced effect,
we can examine the induction of tyrosine aminotransferase in cultured HTC cells (44). These cells, derived
from Morris hepatoma 7288c, retain some but not all of
the responses to glucocorticoids of normal liver. Fig.
3 shows a typical induction curve in these cells. It

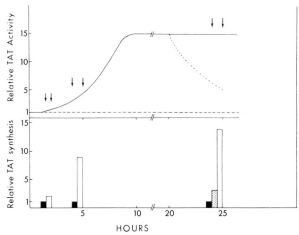

Fig. 3. Rate of synthesis of TAT (lower chart) com-
         pared with kinetics of steroid-provoked in-
         duction (upper chart). Black bars represent
         rate of synthesis in control cells; open bars,
         from induced cells, and dotted bar from cells
         induced and then washed free of inducer. Kin-
         etics of wash-out shown by descending dotted
         curve. Labelled periods with amino acid to
         determine synthetic rates delineated by pairs
         of arrows. (Reproduced with permission from
         ref. 44.)

demonstrates a characteristic kinetic response in enzyme induction by steroids. After inducing steroid is added, there is a delay of 1-2 hr. before new enzyme appears. Analysis of this lag period indicates that there is a 30 min. delay before functional message can be found in the cytoplasm. Since it is known that inducing steroid is in the nucleus within 5 min. or so, there remains an as yet unexplained lag period, perhaps the time required for mRNA processing or release from the nucleus. After the lag is past, enzyme activity rapidly rises to an induced plateau, where it remains unless inducer is removed. This induction in these cells has been shown to be accounted for entirely by increased enzyme synthesis. If inducer is removed, enzyme activity and synthesis fall again to basal levels. Studies on HTC cells have been helpful in clarifying several points about how steroids affect gene expression, points difficult to obtain clear-cut answers to in *in vivo* systems. For example, induction in HTC cells helped demonstrate that the steroid acted directly on the cells it induced, rather than through interactions with other hepatic cells, or release of some intermediate hormone. Since HTC cells do not metabolize steroids, these studies showed that the steroid given, and not some metabolyte, was the inducer. They show that cAMP is probably not required for steroid-mediated enzyme induction since these cells have extremely low levels of cAMP which does not respond to steroids, nor can exogenous cAMP or its analogs evoke steroid-like inductions in HTC cells. The general rRNA and tRNA responses seen in liver after corticosteroids do not occur in HTC cells; therefore these responses are not required for the induction of a specific enzyme. The lack of these and most other *in vivo* responses to inducing steroids shows that these inductions need not be always coordinately controlled. In the case of proteins, one could say that the structural gene was simply lost, but ribosomal and tRNA genes of course are still expressed. Thus the elements controlling the steroid effect on these genes can be separated from those controlling induction of the aminotransferase.

Recently, in collaboration with S. Yang and M. Lippman, we have discovered a further induction in these cells, that of a specific transfer RNA (25,50). We found in whole cells and later in a cell-free system, that with kinetics very similar to those shown by the aminotransferase, steroids induce an increase in phenylalanyl tRNA.

Fig. 4 shows the results of a whole-cell experiment in which dexamethasone was left with the cells for

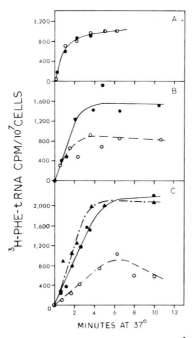

Fig. 4. Kinetics of charging of tRNA^{phe} in HTC cells after varying periods of treatment with dexamethasone phosphate. (A) After 1/2 hr. of steroid ●—●—●; (B) After 1 1/2 hr. of steroid ●—●—●; (C) After 2 3/4 hr. ●—●—●, and 6 1/2 hr. ▲-.-▲.-▲ of steroid. Control cells in all cases o—o—o. (Reproduced with permission from ref. 50.)

varying times, after which radioactive phenylalanine
was added. Samples were taken over the next few minutes, their RNA extracted and its radioactivity estimated. As can be seen, considerably more radioactivity
was found in hormone-treated cells than in control cells.
These treated cells show no overall increase in RNA
synthesis after steroids, and a control experiment
shows that a mixture of labelled amino acids given to
steroid-treated cells does not produce increased labelled protein associated with RNA. The same kind of
results were obtained in a cell-free system with tRNA
charging enzyme from induced or uninduced cells. Re-
verse-phase chromatography of the *in vitro* charged
phetRNA^{phe} showed that the induced increase was in one
of the two isoaccepting peaks (25). This induction is
of potential importance for several reasons. It provides a marker which, together with tyrosine transaminase, allows cell variants to be analyzed for coordinate
control. Since tRNA does not require translation to be
synthesized, it is in a sense a step nearer the genome
than is a protein and may be useful in analyzing the
site of action of steroids. Finally, phetRNA induction
in HTC cells appears to be a rare example of a specific
inducible event which does not occur *in vivo* appearing
in a tumor cell line.

A strong correlation has been made between the
ability of steroids to exert their cell-specific effects
and the presence of steroid receptors in the cytosol.
Steroid receptors represent an opportunity to examine
the detailed molecular events in animal cell gene regulation (28,45). For example, in corticosteroid-inducible
HTC cells Samuels and Tomkins (36) differentiated 4
classes of steroids with respect to their ability to
induce an enzyme: optimal inducers, suboptimal inducers,
non-inducers which compete, and non-inducers which fail
to compete. Any steroid of the first 3 classes would
exhibit competition with other steroids of those classes
in producing the biologic effect. The 4th class seemed
merely inactive. Later, Baxter and Tomkins (7) demonstrated that there was very close correspondence between
the ability of a steroid to bind to the receptor and its

ability to compete for biologic effect. Non-inducing
and non-competing steroids did not bind at all. These
receptors differ from the bulk of cellular sites to
which steroids bind in possessing high affinity and
specificity for the appropriate steroid and in being
saturable at low concentrations of steroid. They appear to be proteins and have been partially purified.
By a differential binding assay, therefore, it is possible to quantitate the number of binding sites, to
estimate the number of classes of sites, and to assess
their affinity for various steroids. It is estimated
that there may be tens of thousands of steroid cytosol
binding sites per cell. They are not steroid-metabol-
izing enzymes. Steroid-receptor binding does not require RNA or protein synthesis. Once bound to receptor,
steroid and protein appear to move to the nucleus (28,
45). This process is not well worked out, but appears
to involve several steps, and in thymocytes at least,
to be sensitive to actinomycin D. What occurs in the
nucleus is not known, but there appear to be several
hundred to several thousand nuclear sites as well.
O'Malley et al. have presented evidence that the steroid-
receptor complex is bound to a specific acid protein
fraction (31). Their experiments with progesterone and
estrogen lead to the proposal that steroid-receptor complex is further restricted as to tissue specificity by
binding to a certain nuclear protein. This interaction,
it is proposed, leads to increased transcription of
specific cistrons. Increased labeling of RNA can be
demonstrated in a cell-free system, but that it includes
specific mRNA is not yet clear.

Baxter and colleagues on the other hand show direct, specific binding of corticosteroid receptor to
DNA (6). They have emphasized the point that ultimately, binding to DNA should depend on specific base sequences in the DNA and their interaction with steroid-
receptor complex. Thus, an intermediate nuclear receptor
protein seems rather redundant. Specificity could be
achieved by simply blocking, presumably with histones
or nonhistone proteins, those recognition sites not appropriate to the tissue. As with estrogens, corticosteroids can cause increased RNA synthesis when allowed

to interact with crude receptor and then to incubate with liver nuclei (8). Again, the technical problem yet to be overcome is to show that the RNA's produced really represent those seen *in vivo*. Otherwise, the entire effect may be misleading.

We have been utilizing cell variants, attempting to obtain clues to the mechanism of steroid control of gene expression. One spontaneous variant cell line, only slightly inducible for tyrosine transaminase, was found to be still fully inducible for surface factor, another known steroid-dependent event (24). Upon examining this line for steroid receptor, we found half the amount in wild-type. These results raise the possibility that there may be multiple receptors each mediating a special effect. Other interpretations are also possible, of course. Other experiments which raise the possibility of various glucocorticoid receptors being responsible for different specific events are those in HTC x L cell hybrids. When we created such hybrids by the technique of somatic cell hybridization, we found several striking results. Induction of tyrosine aminotransferase, characteristic of the HTC parent, was lost. Yet the typical L-cell responses to steroid, partial inhibition of macromolecular synthesis and of glucose uptake, remained. Both the TAT induction and the inhibitions are known to be RNA- and protein-synthesis dependent and to require steroid receptors. When we examined the hybrids for receptors, we found them in an apparently normal state. Further examination, by heat denaturation and by *in vitro* nuclear binding, suggested that L-cell and HTC cell receptors differed and that both were present in the hybrids (25a). It is possible, therefore, that the inductive and inhibitory events of the two cell types are mediated by similar but different receptors, and the type of glucocorticoid receptor within a cell may partially determine the response of that cell to the hormone.

These results, and those with spontaneous variants of the HTC line, show unequivocally that mere presence of receptor does not guarantee responsiveness to steroid.

By cloning the wild-type population in the absence of known selective pressures, we have isolated several low- or non-inducing HTC variants (5). These cells still have apparently normal basal enzyme. Upon examining them for steroid receptor, we found that they had receptor which in quantity and quality appeared like wild-type. Therefore, these cells seem to have a block in their induction apparatus subsequent to receptor, and we hope that they will therefore provide clues to the biochemistry of induction.

So far, everything we have considered is consistent with an action of steroids inducing at the level of transcription. Certain experiments, however, suggest that steroids may be acting through some post-transcriptional control element. One such example is glutamine synthetase in retinal cells. In these cells, *in vivo* and in culture, this enzyme appears at a certain time during development. Administration of hydrocortisone hastens the appearance of the enzyme, an induction which is blocked by inhibiting RNA synthesis or protein synthesis. Nevertheless, actinomycin D, an inhibitor of RNA synthesis, as well as certain other compounds, are capable of inducing the enzyme by itself (30). These experiments have been interpreted to mean that there is a labile post-transcriptional control element which is actinomycin-sensitive.

In HTC cells, we have carried out similar experiments (43), and similar effects have been found in many other systems, *in vivo* and *in vitro* (see references in 43). The simple transcriptional model of hormone action does not account for such results. Fig. 5 illustrates this point. The straightforward transcription model states that the induced level of enzyme depends upon continued transcription of mRNA, due to the presence of inducer. If one either removes inducer or blocks RNA synthesis, enzyme synthesis should decline. We have already seen that removing inducer from these cells results in deinduction (Fig. 3). But Fig. 5 shows that if inducer is removed and actinomycin added, deinduction does not occur. Degradation of enzyme was followed immunologically in this experiment and shown

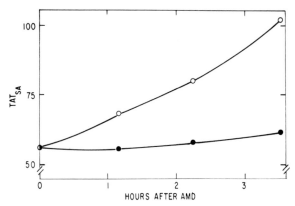

Fig. 5.   Superinduction of TAT by actinomycin D.   21
petri dishes of HTC cells in log growth were
transferred to non-growth medium (Swim's S77
+ 2% dialyzed bovine serum) and were then
made 10^{-5}M dexamethasone phosphate.   16 hr.
later 3 dishes were taken for enzyme assay.
To half of the remainder, 5 µg/ml actinomycin
D was added.   At the times indicated, triplicate dishes were assayed for tyrosine aminotransferase.   Results show averages of the
triplicate assays.   Dexamethasone only,
●—●—●.   Dexamethasone plus actinomycin D,
o—o—o.

to be unaltered by the actinomycin. One interpretation
of this result is that the actinomycin and the steroid
both inhibit a post-transcriptional element which somehow blocks expression of pre-existing messenger RNA
for the transaminase (46). Without going into the de-
bate concerning this model, the most direct test of it
would be to assay directly the content of functional
messenger RNA in the cell under the various conditions
of treatment. Several labs are presently trying to do
such experiments.

Thus the seemingly straightforward transcriptional
system may be more complex after all. Another example
of the complexity of steroid induction lies in the work
on alkaline phosphatase. Hydrocortisone induces this
enzyme in HeLa cells. The induction, as usual, is

prevented by adding actinomycin or cyclohexamide simultaneously with inducer. Radioimmunoprecipitation studies show, however, that the increase in enzyme activity occurs without an increase in the number of alkaline phosphatase molecules. In other words, this enzyme is being activated by some other substance induced by the steroid (12).

There is still another level of control of steroidinducible enzymes which we have not yet discussed. This is the question of why some cells differentiate to respond to the hormone and others do not. The correlation between the presence of steroid receptors and tissue response offers one explanation, but the existence of cell variants containing receptors but which still fail to respond argues that other levels of control may also exist. We have been exploring this question by use of somatic cell hybrids, some of which were mentioned above. This technique consists of fusing two cells together, so that both nuclei exist in a single cytoplasm. Eventually the nuclei fuse as well and a hybrid is formed, with one complete set of chromosomes from each parent. Such fusion products slowly segregate toward a single diploid set of genetic information. Hybrids can be used to study gene control in several ways. They can be used for locating genes on specific chromosomes and for complementation analysis. For instance, Puck and his colleagues have developed a series of auxotrophs in the Chinese hamster ovary cell line and by studying cell hybrids, have demonstrated complementation for the missing functions (22). Hybrids between mouse and human cells have been used in several studies to map human chromosomes (34).

Another way hybrids can be used is to examine dominant/recessive relationships. In independent series of studies we, and Weiss and her colleagues, have shown that fusing a cell inducible for tyrosine aminotransferase with a cell lacking the enzyme invariably results in a noninducible hybrid (38,42). Furthermore, Weiss and Chaplain have shown that with loss of chromosomes, inducibility is sometimes recovered (48).

I will show you just one example of such a study. In this study, we fused the inducible HTC cell with the non-inducible BRL-62 (42). The latter is a diploid epithelial cell line developed from Buffalo rat liver. The HTC cell parent tumor originated in this same in-bred rat strain. We elected to fuse the cells and examine them early after fusion, before the nuclei had an opportunity to join together. To identify true heterokaryons, we labelled the nuclei of one line with ^{3}H thymidine. Thus, we looked for cells with two nuclei, one of them radioactive. Using a histochemical stain, we examined the heterokaryons for transaminase.

    Fig. 6 shows the results of such an experiment. The photomicrograph includes a field of cells, mostly not fused. The smaller HTC cells are deeply stained, while the large BRL-62 cells are not. The fused cell

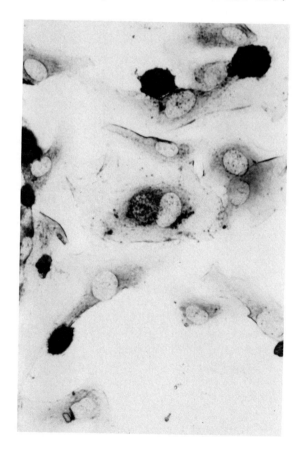

Fig. 6. Combined radioautography and histochemistry
studies to show tyrosine aminotransferase in
a heterokaryon comprised of HTC (TAT$^+$) and
BRL- (TAT$^-$) cells. Fused cell in center,
containing one ^{3}H thymidine-labelled HTC
nucleus (note overlying black silver grains)
and one unlabelled BRL-62 nucleus. Cytoplasm
is histochemically negative for TAT. Sur-
rounding unfused cells show TAT$^-$ BRL-62 cells
and TAT$^+$ HTC cells (intense black reaction)
(Reproduced with permission from ref. 42).

in the center, identified by its containing one ^{3}H
labelled and one unlabelled nucleus, has a level of
staining similar to that of the BRL-62 parent. These
results suggest a rapid dominant effect of the TAT$^-$
parent.

In sum, studies on steroid-mediated cellular events
indicate control at several of the complex levels of
gene regulation theoretically possible. While a great
deal of research has indirectly suggested that steroids
somehow act at a transcriptional level, certain experiments also suggest an element of post-transcriptional
control. Progress in the field of steroid receptors
promises to lead to a clearer understanding of these
events. The combination of biochemical studies, of the
development of appropriate cell-free systems, and of
genetic studies on cell hybrids we hope will assist in
elucidating the molecular events involved in gene regulation in animal cells.

References

1. Anfinsen, C.B. Principles that govern the folding
    of protein chains. Science 181, 223 (1973).

2. Anonymous. The story of poly A. Nature 235, 1
    (1972).

3. Attardi, G., and Amaldi, F. Structure and synthe-
    sis of ribosomal RNA. Ann. Rev. Biochem. 39, 183
    (1970).

4. Attardi, G., Parnas, H., Hwang, M., and Attardi, B. Giant-size rapidly labelled nuclear ribonucleic acid and cytoplasmic messenger ribonucleic acid in immature duck erythrocytes. J. Mol. Biol. 20, 145 (1966).

5. Aviv, D., and Thompson, E.B. Variation in tyrosine aminotransferase induction in HTC cell clones. Science 177, 1201 (1972).

6. Baxter, J.D., Rousseau, G.G., Benson, M.C., Garcea, R.L., Ito, J., and Tomkins, G.M. Role of DNA and specific cytoplasmic receptors in glucocorticoid action. Proc. Nat. Acad. Sci. USA 69, 1892 (1972).

7. Baxter, J.D., and Tomkins, G.M. The relationship between glucocorticoid binding and tyrosine aminotransferase induction in hepatoma tissue culture cells. Proc. Nat. Acad. Sci. USA 65, 709 (1970).

8. Beato, M., Homoki, J., Doenecke, D., and Sekeris, C.E. Effects of ions and cortisol on RNA synthesis in lysed rat liver nuclei. Experientia 26, 1074 (1970).

9. Blatti, S.P., Ingles, C.J., Lindell, T.J., Morris, P.W., Weaver, R.F., Weinberg, F., and Rutter, W.J. Structure and regulatory properties of eukaryotic RNA polymerase. Cold Spr. Harb. Symp. Quant. Biol. 35, 649 (1970).

10. Britten, R.J., and Davidson, E.H. Gene regulation for higher cells: a theory. Science 165, 349 (1969).

11. Brown, D.D., and Littna, E. RNA synthesis during the development of Xenopus laevis, the South African clawed toad. J. Mol. Biol. 8, 669 (1964).

12. Cox, R.P., and Elson, N.A. Hormonal induction of alkaline phosphatase activity by an increase in

catalytic efficiency of the enzyme. J. Mol. Biol. <u>58</u>, 197 (1971).

13. Crick, F. General model for the chromosomes of higher organisms. Nature <u>234</u>, 25 (1971).

14. Dahmus, M.E., and Bonner, J. Nucleoproteins in regulation of gene function. Fed. Proc. <u>29</u>, 1255 (1970).

14a. Darnell, J.E., Jelinek, W.R., and Molloy, G.R. Biogenesis of messenger RNA: genetic regulation in mammalian cells. Science <u>181</u>, 1215 (1973).

15. Donovan, G., and Oliver, I.T. Purification and properties of a microsomal cyclic adenosine monophosphate binding protein required for the release of tyrosine aminotransferase from polysomes. Biochemistry <u>11</u>, 3904 (1972).

16. Georgiev, G.P. On the structural organization of operon and the regulation of RNA synthesis in animal cells. J. Theoret. Biol. <u>25</u>, 473 (1969).

17. Harris, H. The rapidly labelled RNA in the cell nucleus. <u>Nucleus and Cytoplasm</u>, Clarendon Press, Oxford. pp. 49-58, 1968.

18. Harris, H., Sidebottom, E., Grace, D.M., and Bramwell, M.E. The expression of genetic information: a study with hybrid animal cells. J. Cell Sci. <u>4</u>, 499 (1969).

19. Herskowitz, I.H. Regulation of gene action - position effect and dosage compensation. <u>Basic Principles of Molecular Genetics</u>, Little, Brown & Co., Boston. pp. 233-242, 1967.

20. Jacob, S.T., Sajdel, E.M., and Munro, H.N. Different responses of soluble whole nuclear RNA polymerase and soluble nucleolar RNA polymerase

to divalent cations and to inhibition by α-amanitin. Biochem. Biophys. Res. Comm. <u>38</u>, 765 (1970).

21. Johns, E.W. Histones, chromatin structure, and RNA synthesis. Nature New Biol. <u>237</u>, 87 (1972).

22. Kao, F., Johnson, R., and Puck, T. Complementation analysis on virus-fused Chinese hamster cells with nutritional markers. Science <u>164</u>, 312 (1969).

23. Kuntzel, H., and Schafer, K.P. Mitochondrial RNA polymerase from neurospora crassa. Nature New Biol. <u>231</u>, 265 (1971).

24. Levisohn, S.R., and Thompson, E.B. Tyrosine aminotransferase induction regulation variant in tissue culture. Nature New Biol. <u>235</u>, 102 (1972).

25. Lippman, M.E., Yang, S.S., and Thompson, E.B. Transfer RNA in hepatoma tissue culture cells. II. dexamethasone phosphate induction of tRNA^{phe} analyzed under cell-free conditions. Endocrinol., in press.

25a. Lippman, M.E. and Thompson, E.B. Steroid receptors and the mechanism of the specificity of glucocorticoid responsiveness of somatic cell hybrids between hepatoma tissue culture cells and mouse fibroblasts. J. Biol. Chem., in press.

26. Lyon, M.F. Sex chromatin and gene action in the mammalian X-chromosome. Am. J. Hum. Genet. <u>14</u>, 135 (1962).

27. MacGillivray, A.J., Paul, J., and Threlfall, G. Transcriptional regulation in eukaryotic cells. <u>Advances in Cancer Research</u>, Klein, G., Weinhouse, S., and Haddon, A., eds. Academic Press, New York. vol. 15, pp. 93-162, 1972.

28. McKerns, K.W. (ed). The Sex Steroids, Molecular Mechanisms. Appleton-Century-Crafts, New York, pp. 1-454, 1971.

29. Melli, M., and Pemberton, R.E. New method of studying the precursor-product relationship between high molecular weight RNA and messenger RNA. Nature New Biol. 236, 172 (1972).

30. Moscona, A.A., Moscona, M., and Jones, R.E. Induction of glutamine synthetase in embryonic neural retina in vitro by inhibitors of macromolecular synthesis. Biochem. Biophys. Res. Comm. 39, 943 (1970).

31. O'Malley, B.W., Spelsberg, T.C., Schrader, W.T., Chytil, F., and Steggles, A.W. Mechanism of interaction of a hormone-receptor complex with the genome of a eukaryotic target cell. Nature 235, 141 (1972).

32. Packman, S., Aviv, H., Ross, J., and Leder, P. A comparison of globin genes in duck reticulocytes and liver cells. Biochem. Biophys. Res. Comm. 49, 813 (1972).

33. Pastan, I.H. Current directions: research on cyclic AMP. Current Topics in Biochemistry. Anfinsen, C.B., Goldberger, R.F., and Schechter, A.N., eds. Academic Press, New York. pp. 65-100, 1972.

34. Ruddle, F., Chapman, V., Ricciuti, F., Murname, M., Klebe, R., and Khan, P. Linkage relationships of seventeen human gene loci as determined by man-mouse somatic hybrids. Nature New Biol. 232, 69 (1971).

35. Sajdel, E.M., and Jacob, S.T. Mechanism of early effect of hydrocortisone on the transcriptional process: stimulation of the activities of purified

rat liver nucleolar RNA polymerases. Biochem. Biophys. Res. Comm. 45, 707 (1971).

36. Samuels, H.H. and Tomkins, G.M. Relation of steroid structure to enzyme induction in hepatoma tissue culture cells. J. Mol. Biol. 52, 57 (1970).

37. Schimke, R.T. Protein turnover and the regulation of enzyme levels in rat liver. Nat. Cancer Inst. Monogr. 27, 301 (1967).

38. Schneider, J., and Weiss, M. Expression of differentiated functions in hepatoma cell hybrids, I. Tyrosine aminotransferase in hepatoma fibroblast hybrids. Proc. Nat. Acad. Sci. USA 68, 127 (1971).

39. Steven, R.H., and Amos, H. RNA metabolism in HeLa cells at reduced temperature II. Steps in the processing of transfer RNA. J. Cell. Biol. 54, 1 (1972).

40. Suzuki, Y., Gage, L.P., and Brown, D.D. The genes for silk fibroin in Bombyx mori. J. Mol. Biol. 70, 637 (1972).

41. Thompson, E.B. Temporal aspects of macromolecular synthesis in eukaryotic cells. Temporal Aspects of Therapeutics. Urquhart, J. and Yates, F.E., eds. Plenum Publishing Corp., New York, 1973. pp. 71-92.

42. Thompson, E.B., and Gelehrter, T.D. Expression of tyrosine aminotransferase activity in somatic-cell heterokaryons: evidence for negative control of enzyme expression. Proc. Nat. Acad. Sci. USA 68, 2589 (1971).

43. Thompson, E.B., Granner, D.K., and Tomkins, G.M. Superinduction of tyrosine aminotransferase by actinomycin D in rat hepatoma (HTC) cells. J. Mol. Biol. 54, 159 (1970).

44. Thompson, E.B., Levisohn, S.R., and Miller, J.V., Jr. Steroid control of tyrosine aminotransferase in hepatoma tissue culture (HTC) cells. Hormonal Steroids, Proc. 3rd Int. Cong. Excerpta Medica International Congress Series, No. 219, pp. 463-471, 1970.

45. Thompson, E.B., and Lippman, M.E. Mechanism of action of glucocorticoids. Metabolism, Clinical and Exper., in press, February 1974.

46. Tomkins, G.M., Gelehrter, T.D., Granner, D., Martin, D., Jr., Samuels, H.H., and Thompson, E.B. Control of specific gene expression in higher organisms. Science 166, 1474 (1969).

47. Wall, R. and Darnell, J.E. Presence of cell and virus specific sequences in the same molecules of nuclear RNA from virus transformed cells. Nature New Biol. 232, 73 (1971).

48. Weiss, M. and Chaplain, M. Expression of differentiated functions in hepatoma cell hybrids: reappearance of tyrosine aminotransferase inducibility after the loss of chromosomes. Proc. Nat. Acad. Sci. USA 68, 3026 (1971).

49. Williamson, R., Drewienskiewicz, C.E., and Paul, J. Globin messenger sequences in high molecular weight RNA from embryonic mouse liver. Nature New Biol. 241, 66 (1973).

50. Yang, S.S., Lippman, M.E., and Thompson, E.B. Transfer ribonucleic acid in hepatoma tissue culture cells. I. Induction in vivo by dexamethasone phosphate of phenylalanine-accepting activity. Endocrinol., in press.

PLASMA LIPOPROTEINS AND APOLIPOPROTEINS

Donald S. Fredrickson

Molecular Disease Branch
National Heart and Lung Institute
National Institutes of Health
Bethesda, Maryland 20014

Given a quantity of plasma (or serum), some salt to raise its density, and an ultracentrifuge, one can isolate fairly easily the lipoproteins from all the other plasma proteins. The plasma lipoproteins contain nearly all the fats and lipids and a very small quantity (2-4 per cent) of the total proteins in plasma.

The lipoproteins have developed in higher forms of life for the inter-organ transportation of lipids, substances of little solubility in water. The lipoproteins provide dispersions of lipids that are at once stable and yet easily handled at their destinations. Critical elements in this packaging for export are certain highly specialized proteins or apolipoproteins.

Homologues of the apolipoproteins in man appear much earlier in the evolutionary scale and they are today subjects of considerable interest. They are important to those who study fat transport as an entree into the causes of certain human diseases. And they are of interest to many as models of important interactions between lipids and proteins that occur within cells.

It is my intent here to draw together some of the recently acquired knowledge about plasma apolipoproteins. It is still quite inadequate but does permit some reconstruction of their properties and functions and a forecast of the direction of further research.

The Spectrum and Structure of Lipoproteins

Note in Table 1 the principal lipid components in plasma and the four classes into which their carrier

Table 1 - The principal lipids and lipoprotein classes in plasma.

Lipids	Concentration[*] in mg/100 ml
Phospholipids	230
Cholesteryl Esters	130
Tryglycerides	100
Cholesterol (unesterified)	60
Lipoprotein classes	
Chylomicrons	0-25
Very Low Density Lipoproteins (VLDL)	120
Low Density Lipoproteins (LDL)	400
High Density Lipoproteins (HDL)	300

[*]Approximate concentrations in plasma withdrawn from a healthy young adult male after an overnight fast. The triglyceride, chylomicron and VLDL concentrations will vary considerably during the diurenal cycle.

lipoproteins are separated according to differences in density or electrophoretic mobility. The latter is dependent upon both net charge and particle size. The common abbreviations for lipoprotein classes given in Table 1 will be used forthwith in the text. Something of the size range of lipoproteins can be seen in Fig. 1. The largest particles are represented by chylomicrons which are from 1000 Å to about 5,000 Å in diameter. The separation between small chylomicrons and large VLDL (600 to 1,000 Å) is indistinct and their definition is mainly a functional one: chylomicrons are considered the principal bearers of dietary triglyceride,

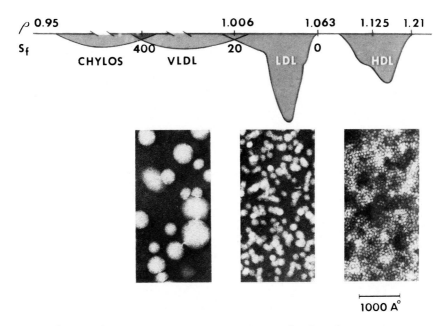

1000 Å

Fig. 1. Schematic representation of the four major
         plasma lipoprotein families, visualized as
         Schlieren patterns obtained in the analytical
         ultracentrifuge, and the appearance of nega-
         tively stained VLDL, LDL and HDL under the
         electron microscope. (Micrographs courtesy
         of Dr. R.W. Mahley.)

and VLDL, of triglycerides arising endogenously. Normally, the bulk of the cholesterol and phospholipids
resides in the LDL and HDL. Each of the lipoprotein
families represents a continuum in terms of particle
sizes and composition and there is considerable heterogeneity within a given broad density class.
     Today it is impossible to describe the true struc-
ture of a plasma lipoprotein. Somewhat more success
has been achieved in creating models of certain membranes. Both plasma lipoproteins and membranes involve
similar kinds of intermolecular arrangements. Lipids
interact with proteins such that a stable oil-water
interface is maintained. An "oil-phase" exists which

221

contains non-polar molecules such as cholesterol, cholesteryl esters, and triglycerides; into this phase
also extend the hydrophobic portions of phospholipids,
molecules which contain both polar and non-polar regions.
The fatty acids in the esterified lipids, being both saturated and unsaturated, have different melting points;
at blood temperature the lipids represent a liquid crystalline state. The apoproteins are inherently more
polar than lipids, but undoubtedly contain both hydrophobic and hydrophilic surfaces. Some of the protein is
at the surface of the oil droplet or liquid crystal and
quite probably other portions of the molecule extend
into the lipid region. As we shall see, the hydrophobic
regions of apoproteins may not be immediately apparent
in their primary structure, and it is possible they depend upon the secondary and tertiary structure of the
apoprotein.

Lipids and protein are not covalently bound in
plasma lipoproteins. Cohesion is achieved mainly
through hydrophobic bonds, forces which derive their
strength from close packing of non-polar regions and
are enhanced when these regions extend for the considerable lengths (20-30 Å), represented by the alkyl
chain of a fatty acid or sphingosine base, or the extended plane of a cholesterol molecule. Electrostatic
interactions are possible between amphipathic heads of
the two principal phospholipids in plasma lipoproteins,
phosphatidyl choline (lecithin) and sphingomyelin, and
free carboxylic or amino groups on the protein chain.
It is noteworthy that two of the known apoproteins,
shortly to be introduced as apoproteins B and C-III,
contain glycosidic residues, including sialic acid,
hexose and hexosamine. It remains to be determined
what importance these sites have in the structure and
metabolism of the lipoproteins.

At the primitive level, one can describe two basic
models for plasma lipoproteins, sketched in Fig. 2.
They are the micelle and the bilayer; some lipoproteins,
like VLDL, may prove to contain features of both.
There is some evidence that HDL are micellar in structure; this is based on X-ray diffraction patterns obtained by Shipley, Atkinson, and Scanu. The HDL sphere

A   B

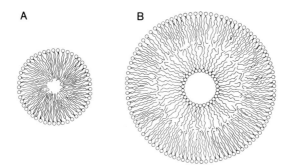

Fig. 2.   Two possible basic models applicable to
          plasma lipoproteins: the micelle (A) and
          the bilayer (B).

shown in Fig. 2, A, is designed to show its diameter
spanned by two extended phosphatidyl choline molecules
(each about 35 Å). This is consistent with the diameter of smaller HDL molecules, which is 80 to 100 Å.
The spacing or relative placement of sterol and steryl
ester molecules among the phospholipids, and in what
manner protein segments may extend into the lipid phase,
cannot be predicted at this time.
     The lipid bilayer (Fig. 2,B), a suggested structure
for some cellular membranes, is another possible model.
In this instance, the spherical particle would contain
a layer of phospholipid and possibly protein at the
center. It has been suggested by Mateau and coworkers,
again from X-ray scattering data, that LDL has a bilayer
structure. In addition to optical techniques for measuring changes in conformation of apoproteins, or pieces
of them, as they are exposed to lipid, the current use
of special probes shows promise of further enlighten-
ment of lipoprotein structure. These include lipid
facsimiles containing free radicals whose electron spin
resonance indicates the constraints placed upon them
when they are inserted into lipoproteins or combined
with apoproteins. More promising are the newer adap-
tations of instruments for measuring nuclear or proton
magnetic resonance which will permit recognition of
the bonding between apoproteins and natural lipids
containing ^{31}P and ^{13}C. The enrichment of ^{13}C content

223

of specific carbon atoms in the lipid molecules should permit their interactions with adjacent structures to be localized in considerable detail.

I would like to leave the subject of the quaternary structure of lipoprotein complexes now, except to emphasize one other important relationship of the apoproteins to the overall structure of lipoproteins. In addition to the role of apolipoproteins as stabilizer or emulsifier of complex mixtures of lipids, they probably play some part in regulating the activity of enzymes that decompose lipoprotein lipids. Most of these enzymes are hydrolases that must be peculiarly adapted to deal with a substrate available to it only at an oil-water interface. In discussing such enzymes and what he calls their micellar "supersubstrates," Brockerhoff has pointed out the requirement for sites on the enzyme, in addition to the catalytic one, where activity is likely regulated by interactions with other lipids and proteins of widely differing polarity. The rate of reaction thus will be dependent on the overall charge, size, and shape of the "supersubstrate." This is well illustrated by the hydrolase, lipoprotein lipase, which catalyzes the hydrolysis of ester bonds between glycerol and long-chain fatty acids. It has long been known that lipoprotein lipase will not act upon triglyceride emulsions unless a plasma cofactor is present. The latter has turned out to be a combination of phospholipid and an apolipoprotein, constituents of the micelles in which triglycerides circulate in plasma.

The Secretion and Transport of Triglycerides. Very Low Density Lipoproteins.

At birth, the plasma lipoprotein pattern in man is like that of fasting adults in most other mammalian species. There is mainly HDL, lesser amounts of LDL, and little VLDL. Within hours after birth, there is a rapid transformation. HDL concentrations rise some, but VLDL, and especially LDL, greatly increase. Transiently, after fat feedings, chylomicrons also appear. A portion of the marked postnatal rise in LDL concentrations is regulated by the sterol and (saturated) fatty acids in the infant diet, but the major contributor is probably the sharp increase in triglyceride transport. Triglycerides represent the bulk of cargo transferred by lipoproteins through plasma in the course of a day.

Triglycerides are found mainly in chylomicrons and VLDL. These lipoproteins include all which do not sediment upon ultracentrifugation at density 1.006. They are particles of a minimum diameter of about 300 Å and may be $> 5,000$ Å. The two classes of particles are arbitrarily separated in the ultracentrifuge by differences in flotation rate, the traditional boundary between VLDL and chylomicrons being S_f 400. (The S_f may be thought of as a negative Svedberg or sedimentation unit, based on the rates of flotation of lipoproteins at 26° and density 1.063.) Chylomicrons and VLDL also can be partly distinguished by electrophoretic migration and flocculation techniques. The physiologist's operational distinction is to take all fat away from the diet of a donor for a few days. The resulting plasma or lymph lipoproteins having density > 1.006 are considered to be transporting only endogenous glycerides and are called VLDL.

VLDL first become visible in the endoplasmic reticulum and Golgi apparatus of the hepatocyte and epithelial cell of the small intestine. Lipid-rich particles, having a diameter of 300-1000 Å, are shown filling the tubules and vesicles of the hepatic Golgi body in Fig. 3. These are VLDL, or their most proximate precursors, that shortly will be secreted into the circulation. They have been assembled to remove triglycerides that are accumulating as the result of net synthesis controlled by a number of operators. One is the metabolism of free fatty acids (FFA). When FFA are delivered to the liver in amounts greater than the capacity for their oxidation, the excess is mainly returned to adipose tissue storage sites as triglycerides. Glucose, in excess of that required to meet oxidative demand or to maintain the hepatic supply of glycogen, is also converted to triglycerides and similarly transshipped. VLDL therefore represent an important system for maintenance of caloric homeostasis. Within the Golgi apparatus of

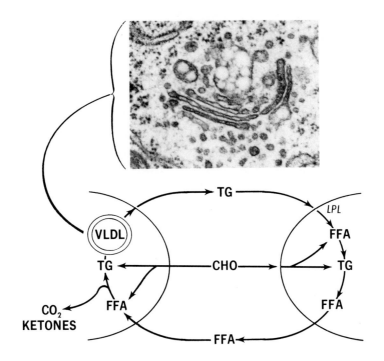

ENDOGENOUS TRIGLYCERIDE METABOLISM

Fig. 3.   Schematic representation of some elements
related to endogenous triglyceride metabolism.
Abbreviations: TG, triglycerides; FFA, free
fatty acids; LPL, lipoprotein lipase; CHO,
carbohydrate. (Photograph of VLDL in liver
Golgi apparatus courtesy of Dr. R.W. Mahley.)

intestinal epithelial cells, two populations of differ-
ent-sized lipoproteins are sometimes visible. Smaller
VLDL particles stand apart from larger droplets. These
latter are chylomicrons that will bear dietary glycerides into the intestinal lymph and from there to the
systemic circulation.

In addition to cholesterol and phospholipids, the
packing of triglycerides for transport includes at
least a half-dozen apoproteins. Because collecting
Golgi bodies is hard work, most of the information about

the particles has come from plasma VLDL. Some valuable insights have also come from VLDL collected in perfusates of liver or intestine.

Plasma VLDL. Plasma VLDL (Fig. 1) bear strong physical and chemical resemblance to particles in the Golgi apparatus, but perhaps we should insert a caveat at this point. Most of the "normal VLDL" studied in man has come from hyperlipoproteinemic patients who have something wrong with delivery or removal of lipoproteins. They may have contributed abnormal VLDL to the pools and the "standards" may be biased. Chemical differences in apoproteins from different donors have been reported and we have seen much variation in their quantity. It will take time to straighten out real differences from artefacts of analysis. VLDL and other lipoproteins captured for chemical characterization also are usually "washed" by repeated ultracentrifugation. Selective losses occur and the final composition may deviate significantly from that of the native micelles in the circulating blood.

Negatively stained VLDL appear spherical in electron micrographs (Fig. 1), and vary greatly in size. This is in accord with their representing a collection of subpopulations that have flotation rates varying from S_f 20 to > 400 and estimated molecular weights that differ by millions of daltons. The "typical" composition shown in Fig. 4 is only an average for the collective content of all VLDL, *i.e.*, all the "endogenous" plasma lipoproteins that do not sediment at density 1.006. The triglyceride:protein ratio (w/w) in the larger VLDL particles (S_f 20-60) is about 10:1 and decreases to 3:1 in the smallest particles (S_f 20-60). The proportions of phospholipid to protein and of free cholesterol to phospholipid are more constant. This prompts one to imagine VLDL as rather expansible "sacks" in which variable amounts of triglycerides are stabilized by proteins and other lipids. The internal proportions of these stabilizers are apparently more important than their absolute quantity. The apolipoproteins present in VLDL are represented in Fig. 4. Their proportions also vary from one end of the VLDL

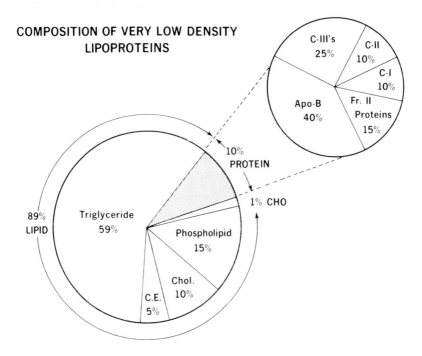

Fig. 4. Composition of VLDL expressed in percentage of anhydrous weight. For fuller explanation, see text. Abbreviations: C.E., cholesteryl esters; chol, unesterified cholesterol; CHO, carbohydrate.

spectrum to the other. Before we focus on these proteins, I need to say something about how they are called.

Apoprotein Nomenclature. The literature contains several different names for most of the apoproteins (Table 2). For a considerable time we identified them by their carboxyl terminal residues, as apoLP-glu, apo-glu, R-glu, etc., because this was one of the first chemical characteristics to be determined as specific proteins were isolated. This system has had its problems; mistakes have been made in assignments and several apoproteins have turned up with the same terminal residues. As a final straw--and as might have been expected--it has recently been shown that apoproteins in other

Table 2 - Nomenclature of plasma apolipoproteins.

A-I	apo-Gln-I, R-Gln-I, (R-Thr), Fraction III
A-II	apo-Gln-II, R-Gln-II, Fraction IV.
B	apo-LDL
C-I	apo-Ser, (apo-Val), D-1, Fraction V
C-II	apo-Glu, D-2, Fraction V
C-III-1	apo-Ala$_1$, D-3, Fraction V, (mono-sialylated C-III)
C-III-2	apo-Ala$_2$, D-4, Fraction V, (di-sialylated C-III)
R-X$_1$	a Fraction-II protein (VLDL); "Arg-rich peptide"
R-Y$_1$	a Fraction II protein (VLDL)

species have terminal residues that are different from their homologues in man. To minimize further confusion, I think it is better to adopt some simpler, more arbitrary designations for the present.

For the reasonably well established apoproteins, we will temporarily employ the system suggested by Alaupovic and coworkers who identify A, B, and C apoproteins, more or less as they have been discovered chronologically. Support is lacking for certain assumptions on which this system was proposed, *i.e.*, that the A-proteins are part of one complex and the C proteins integrated units in another, or that A-I and A-II are more closely related than either might be to one or more of the C-proteins. Either or both of these assumptions may eventually prove to be true. The more valid objection to adoption of the ABC system now is that it does not provide us with any basis for dealing with new apoproteins as they are discovered. The difficulties arising from premature adoption of a classification for substances of uncertain functional relationships have been amply demonstrated in the fields of blood-clotting or the complement system. I will here use the ABC codes where they are generally accepted and refer to one or two newcomers as they were described by their discoverers.

ApoVLDL

Human VLDL contains four well accepted apoproteins and several less well characterized. Their separation as represented in Figures 5 and 6 is similar to that

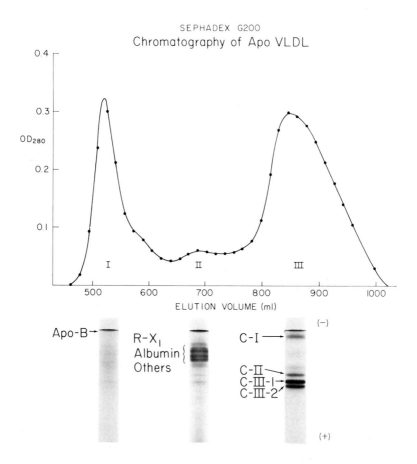

Fig. 5. Preliminary separation of delipidated VLDL apoproteins from human plasma. The contents of Fractions I, II and III are illustrated in

the polyacrylamide gels displayed below them. In this and subsequent figures, the electrophoreses were carried out in 10% gels containing urea, at pH 9.4. Techniques as described by Brown *et al.*: J. Biol. Chem. <u>244</u>: 5687, 1969.

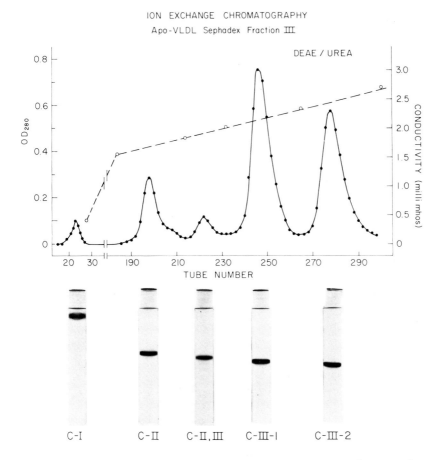

ION EXCHANGE CHROMATOGRAPHY
Apo-VLDL Sephadex Fraction III

C-I   C-II   C-II,III   C-III-1   C-III-2

Fig. 6.   Purification of VLDL Fraction III (Fig. 5) by DEAE chromatography. Polyacrylamide gels run as in Fig. 5.

employed by Brown in our laboratory when he first isolated the C-proteins. (In those days we called them the "D-peptides," a name which persists in some of the

literature. They are proteins, not peptides, and "D"
has given way to "C" or to the carboxyl terminal designations, Table 2). Upon gel filtration of delipidated
VLDL protein the first fraction obtained contains apoB,
the characteristic and nearly sole apoprotein of LDL.
The second minor peak, Fraction II, is potentially important but incompletely characterized. The third
fraction contains the C-proteins. These latter have
relatively low molecular weights and can be separated
into five major peaks on ion exchange chromatography
(Fig. 6).

ApoC-I. The first protein separated from among
the C-protein group by DEAE chromatography is C-I, a
protein having a molecular weight of 6,550. Its carboxyl-terminal residue is serine (it is given as valine
in the earlier literature) and threonine is the amino-
terminal residue. C-I contains a high content of lysine
(16 mols per cent) (Table 3), which makes it highly soluble at acid pH and barely mobile in alkaline polyacrylamide gel electrophoresis (Fig. 6). It contains no histidine, tyrosine, or cysteine (Table 3) and no carbohydrate.

ApoC-III. The last two peaks from the DEAE column
(Fig. 6) contain a protein having alanine as the carboxyl terminal residue. Its polymorphism is due to
differences in sialic acid content. ApoC-III-1 contains
one mol of sialic acid per mol of protein and apoC-III-2,
which migrates faster on polyacrylamide, contains two
mols of sialic acid per mol of protein. Isolation of
an asialo form (C-III-0) has also been reported. The
sialic acid is probably the terminal member of a short
glycosidic chain that also contains equimolar amounts
of galactose and galactosamine and is attached to threonine six residues removed from the carboxyl-terminal
amino acid. C-III contains no cysteine; we find no
isoleucine, although others have described a "polymorphic form" containing this amino acid. Its anhydrous
molecular weight is 8764. C-III is the most abundant
C-protein in VLDL and is present along with the other
two C-proteins in HDL. When all its polymorphic forms
are taken together, C-III constitutes the fourth most
abundant apolipoprotein in human plasma.

232

Table 3. Amino acid composition of the apolipoproteins

	A-I	A-II	B	C-I	C-II	C-III
			mols per cent			
Lys	9	12	8	16	7	8
His	2	0	2	0	0	1
Arg	7	0	3	5	1	3
Asp/Asn	9	4	11	9	7	9
Thr	4	8	7	5	10	6
Ser	6	8	9	12	11	14
Glu/Gln	20	19	12	16	17	13
Pro	4	5	5	2	5	3
Gly	4	4	5	2	3	4
Ala	8	6	6	5	8	12
1/2 Cys	0	1	1	0	0	0
Val	5	8	5	4	5	8
Met	1	1	2	2	2	3
Ile	0	1	5	5	1	0
Leu	17	10	12	11	10	6
Tyr	3	5	3	0	6	3
Phe	3	5	5	5	3	3
Tryp.	2	0	N.D.	2	3	4
PCA	—	1	—	—	—	—
Glu/Gln	—	8/7	—	7/2	—	5/5
Asp/Asn	—	2/1	—	4/1	—	7/0
M.W.	28,000	17,380	?	6,550	9,000	8,764

Primary and Secondary Structure

The first of the apolipoproteins for which amino acid sequences were determined were C-I and C-III. You will note in Fig. 7 a paucity of segments rich in aromatic or other less polar residues that are obvious

233

Apolp-Ser or C-I

Apolp-Ala or C-III

Apolp-Gln-II or A-II

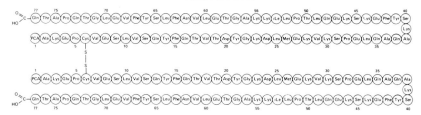

Fig. 7. The amino acid sequence of human C-I (R. Shul-
man *et al*.: Circulation <u>XLV</u>: II-246, 1972);
C-III (H.B. Brewer, Jr. *et al*.: Adv. Expt.
Biol. Med. <u>26</u>: 280, 1972); and H-II (Brewer
et al.: Proc. Nat. Acad. Sci. USA <u>69</u>: 1304,
1972).

sites of hydrophobic bonding with lipids, presumably
the raison d'etre of an apolipoprotein. As mentioned
earlier, however, the presentation of hydrophilic and
hydrophobic areas must be partially dependent upon the
conformation of the proteins in their natural environment. Both C-I and C-III have an appreciable amount of
alpha-helical structure, and this is considerably increased when phospholipid is recombined with the delipidated apolipoprotein.

There is another feature of the chains that may be
of importance. There are a number of acid-base pairs,
viz. adjacent combinations of Lys and Glu. These oppo-
sitely charged molecules could form points of attachments
for the amphipathic polar head groups of phospholipids.
Actually, the number of acid-base conjunctions in either

234

C-I or C-III is not significantly different from that expected from random distribution within chains having their amino acid compositions. Acid-base pairing of the degree observed is also not unique to apolipoproteins.

Genetic Relationship Between C-I and C-III

The primary structures of proteins sometimes permit comparisons that shed light on their relationships and evolution. Barker and Dayhoff (Table 4) have examined

Table 4. Relationships of human apolipoproteins

Sequence 1	Sequence 2	Alignment Score	p
C-I	C-III (1-59)	4.56	10^{-5}
C-I	A-II (1-59)	4.66	10^{-5}
A-I (1-59)	C-III (1-59)	3.66	10^{-3}
A-II	C-III	3.46	10^{-3}

Alignment scores (in S.D. units) were obtained using the mutation data matrix of Barker and Dayhoff for comparison of complete sequence of apo-Ser and shortened or complete sequences of apo-Ala and apo-Gln-II. The p value (derived from 270 random runs in each comparison) is the calculated probability that the similarities in amino acid sequence would occur by chance. (Data kindly supplied by Dr. Winona Barker.)

the sequences of several apolipoproteins, using a "mutation matrix," which scores the amino acid alignments of different chains, taking into account the nature of the genetic code, the rates of mutation at the nucleotide level, and natural selection. The method determines the highest score for any alignment (including gaps) of two protein sequences. This score is then compared with the highest possible scores obtained by aligning pairs of randomized sequences having the same amino acid composition as the two real sequences. The alignment score is

expressed in units of standard deviation from the mean
or random scores.

By this technique, C-I and the first portion (residues #1-59) of C-III appear to have arisen from a common
ancestral gene. Possibly they began as proteins of similar length; and, during their divergent evolution,
C-III was elongated and acquired a glycosidyl side
chain. The alignment score for C-I and the last 59
residues of C-III (apo-Ala) is not different from that
expected from random distribution.

Thus, these two apoproteins, considered in the
scheme of Alaupovic and coworkers to be parts of a
C-protein complex, do have a probable genetic relationship consistent with their appearance together in common portions of the density spectrum of plasma lipoproteins. This does not prove, of course, that their
synthesis and catabolism are inseparable or that they
function only as a unit within the structure of lipoproteins.

C-II

The primary structure of C-II (molecular weight
~ 9,000) (Fig. 6) is not yet known. We would like to
know it especially because a special function has been
found for C-II, which I will mention later. C-II contains no cysteine or histidine and is relatively rich
in serine and glutamic acid/glutamine residues (Table
3). Optical measurements of lipid-free C-II do not
suggest much ordered secondary structure, at least in
the lipid-free state.

ApoB (ApoLDL)

The largest protein component that is consistently
isolated from the VLDL, and is also an important protein in chylomicrons, is apoB. You will recall it in
Fig. 5 as the first or void volume peak emerging from
Sephadex loaded with delipidated VLDL proteins. It
does not enter the polyacrylamide gels shown in Fig. 5.
ApoB is also called apoLDL, for it is almost the sole
protein constituent of the S_f 0-12 subclass of LDL

family. In adult human males the concentration of LDL alone is sufficient to make the concentration of B greater than that of any other apoprotein in plasma.

Its resistance to characterization, its seeming essentiality for glyceride transport, and perhaps the added suspicion that it has something to do with atherogenesis, have all transformed apoB into one of the central mysteries of lipoprotein physiology. Questions of its size and homogeneity remain unanswered. Tanford and coworkers have concluded that B is a single protein having a minimal molecular weight of about 300,000. Still others conclude it may be a heterogeneous complex of several proteins. ApoB is a glycoprotein containing sialic acid.

Apolipoprotein B aggregates and becomes insoluble as soon as the lipids have been removed from it. It can be re-dissolved only at high pH or when disaggregating agents such as guanidine or detergents are added. The unmanageability of delipidated apoB is reminiscent of other membrane proteins and leads one to ponder again the bilayer model of LDL proposed by Mateau in Fig. 2. It is conceivable that this lipoprotein is a portion of a membrane spewed forth from the liver or intestine in the process of triglyceride secretion. In circulating LDL apoB is complexed with a fairly constant mixture of cholesteryl esters, free sterol, and phospholipids. The latter include phosphatidyl choline, and sphingomyelin, in a molar ratio not far from 1. In VLDL and other lipoproteins containing large amounts of glycerides, however, the membrane-like LDL "core" is temporarily associated with both a great deal more non-polar lipid and some additional protein. These other proteins include the C-proteins, and as we shall see presently, there are reasons to believe that some or all of them would most likely be located at the periphery of the triglyceride-rich particle.

Other VLDL Apoproteins. In Fig. 4 about 15 per cent of the VLDL complement of apoprotein is assigned to the "Fraction II" eluted from Sephadex between apoB (Fraction I) and the C-proteins (Fraction III). They

behave as though they have molecular weights of 25,000
to 50,000 daltons, but none has been completely characterized. A prominent member, first described by
Shore and Shore as "R-X$_1$," is unusually rich in arginine and glutamic acid. A companion, "R-X$_2$," contains
high proportions of serine, glycine, and glutamic acid.

ApoVLDL in Other Species. Plasma VLDL from the
rat and dog have a complement of apoproteins very similar to that in man (Fig. 8). All contain apoB as a

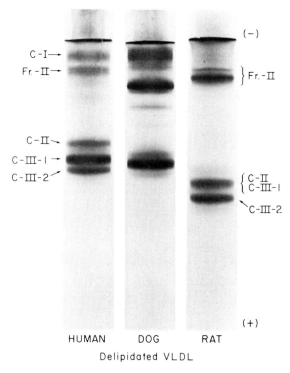

HUMAN   DOG   RAT

Delipidated VLDL

Fig. 8. Polyacrylamide gel patterns of VLDL apoproteins in man, dog, and the rat. In the rat,
the equivalent of C-I is not stained well,
but is in the same position as in man.

major fraction; the equivalents of C-I, C-II, and several forms of C-III are there, as are "R-X$_1$" and other
"Fraction II" bands. ApoB and some of the C-proteins

also have been identified in VLDL obtained from rat liver Golgi apparatus and in "liposomes" obtained from rat liver.

Apoprotein Synthesis

The apoprotein content of lipid droplets in cells secreting triglyceride has led to the conclusion that they are VLDL in a "prenatal" state. VLDL have been collected from perfused, isolated liver and in the lymph from small intestine. These lipoproteins also contain the B and C-proteins common to the VLDL usually present in plasma. Something else has been learned, however, by adding labelled amino acids to the perfusates, thus permitting some measure of new apoprotein synthesis during triglyceride secretion. Under these conditions, B and C-proteins, and the A proteins found in HDL that I have yet to describe, all seem to be made in the isolated liver. During VLDL formation, however, the liver synthesizes apo B and the "Fraction II" proteins far more actively than it synthesizes C-proteins. Moreover, the perfused rat small intestine produces VLDL that contains newly synthesized apo B and "Fraction II" apoprotein, but the C-proteins do not appear to be made at this site. In the rat, at least, the VLDL secreted into intestinal lymph thus appear to obtain their complement of C-I, -II, and -III from other lipoproteins circulating in the lymph after the VLDL emerge from the cell.

Apoproteins in Metabolism of VLDL

Indirect evidence allows us to guess at important roles for two of the several apoproteins that are either secreted with VLDL or quickly become intimately associated with them in plasma.

VLDL Synthesis: Essentiality of ApoB

ApoB has been the subject of many clinical or physiological experiments. LDL is a powerful antigen

and provokes antibodies that are directed to its principal apoprotein without cross-reaction with any of
the other apoproteins. Thus there exist sensitive immunochemical techniques for detecting the presence of
apoB in body fluids and tissues. These have been used
in the rare human mutant, abetalipoproteinemia, in attempts to ascertain the relative importance of different apoproteins in enabling the secretion of triglycerides from the intestine and liver.

In abetalipoproteinemia there is a life-long absence of plasma chylomicrons, VLDL, and LDL. Only
HDL remain in circulation. The triglyceride content
of plasma is thus vanishingly small; ingested glycerides stagnate in the villi of the small intestine until they are eventually taken up through the portal
circulation. The HDL remaining in the plasma contain
C-I, C-II, C-III-2 (the more heavily sialylated ver-
sion of C-III), and the two major HDL apoproteins,
A-I and A-II. We have not been able to detect a trace
of immunochemically reactive apoB and have concluded
that it is either not synthesized or utilized in
abetalipoproteinemia. This supports, but does not
prove, the hypothesis that apoB must be released if
triglyceride is secreted. In the perfusion experiments the vague Fraction-II proteins also seemed to be
turning over briskly as VLDL emerge from either liver
or gut, and the essentiality of one or more of these
proteins must also be tested. There are other possible explanations. The liver in the patients with
abetalipoproteinemia is heavy-laden with triglyceride.
When plasma is perfused as a source of LDL and apoB,
it does not stimulate any triglyceride release. But
the interpretation of this experiment hinges on the
unanswered question of whether apoB is re-utilizable
by liver or any other organ. In fact, the tissue sites
of disposal of such an important protein are not established. Thus, it is in a tentative way that we
think of apoB as having to be elaborated anew with
intracellular formation of each triglyceride-rich
particle.

VLDL Catabolism: Triglyceride Hydrolysis

The mechanisms whereby the triglyceride-rich particles are metabolized are the subject of a literature
rich in both detail and uncertainty. We begin with
the unproved assumption that the glycerides in chylomicrons and VLDL are removed by similar mechanisms.
The rate seems to be directly related to the size of
the particles. As the particles course through the
capillary beds, their triglycerides are subject to attack by lipase(s) at the surface of endothelial and
possibly hepatic cells. Ester bonds are hydrolyzed and
some products, including partial glycerides, are passed
beyond into the tissues. Some of the free fatty acids
and glycerol, and "remnant lipoproteins," which are
poorer in triglyceride content and altered in both
lipid and apoprotein composition, continue on in the
blood to other fates.

The principal enzyme catalyzing hydrolysis of
circulating triglyceride is lipoprotein lipase (LPL)
(EC 3.1.1.3). It is obtainable in high concentration
in extracts of adipose tissue and blood vessels. LPL
appears to be bound to elements near the capillary
membrane, for it is displaced into plasma when heparin
is given. Several members of our laboratory, particularly LaRosa and Krauss, have devoted years of work to
prove that plasma post-lipolytic activity includes in
addition to LPL, similar but non-identical lipase activity that is derived from the plasma membrane of the
liver. Korn showed many years ago that lipase extracted
from adipose tissue was not able to hydrolyze triglyceride in artificial emulsions unless a small amount
of plasma or plasma lipoproteins was present. Recently
we and Havel and coworkers have independently demonstrated that this requirement for lipoprotein lipase
"activation" can be fulfilled by apoC-II, provided
phospholipid is also present. A typical experiment
showing the enhancement of lipolysis by this apoprotein
is shown in Fig. 9.

These activation experiments are conducted by mix-
ing labeled triolein, phosphatidyl choline, the apoprotein, and enzyme extracted from adipose tissue, the

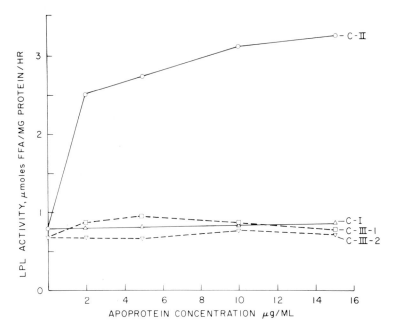

Fig. 9. The effect of various apoproteins on the
hydrolysis of ^{14}C-triolein by adipose tissue
extracts of lipoprotein lipase. Methodology
as described by R. Krauss *et al*.: Circ. Res.
XXXIII: 403, 1973.

mixture being dispersed by sonication just prior to incubation. All the pure apoproteins available have
been similarly tested for effect on LPL activity. The
results have varied in different laboratories. One has
reported that C-I is also an activator of partially
purified plasma LPL; others have concluded, on the one
hand, that C-III is an activator, or a specific inhibitor, on the other. We have found that certain concentrations of all the other apoproteins can oppose the
enhancement of hydrolysis by C-II. Such disparity is
not too surprising from the consideration we mentioned
earlier about lipid hydrolases and their need to ac-
commodate to a "supersubstrate" involving an oil-water
interface. The chylomicron or VLDL is not precisely
duplicated in these *in vitro* experiments. Sterol and
sphingomyelin, and certain other proteins are missing

from the micelles; and "regulation" of enzyme-substrate interaction by one or another apoprotein that is observed under the artificial conditions is not necessarily occurring *in vivo*. Nevertheless, it is likely that, when purified LPL and C-II become available, interactions between them and certain lipids will be discerned that will explain more specifically how hydrolysis of glyceride is regulated by proteins other than enzymes at the surface of the native micelle.

The several heparin-displaced lipase activities are so similar that it has not been excluded that they may contain the same protein subunit. Nevertheless, the lipase activity that is displaced by heparin from the liver plasma membrane can be distinguished from LPL in assays based on differences in sensitivity to ionic strength or to protamine. The hepatic lipase is not affected by any of the apoproteins *in vitro*. The function of this enzyme is not known. Experiments suggest, however, that it prefers smaller VLDL rather than chylomicrons as substrate. Thus LPL, and its activator(s), would seem to be more important in the early stages of metabolism of large triglyceride-bearing lipoproteins; other lipases, including the hepatic enzyme, may be operating primarily on smaller remnants.

There are several genetic disorders in man characterized by massive hypertriglyceridemia. In one (type I hyperlipoproteinemia) chylomicron accumulation is so severe that the plasma usually has the appearance and consistency of heavy cream. The LPL activity in adipose tissue of these patients is very low, although the total post-heparin lipolytic activity sometimes is near normal. The new method of Krauss, which differentiates between LPL and the hepatic lipase in post-heparin plasma, provides a simple test for highlighting the LPL deficiency. The hepatic lipase activity is normal. Other hyperchylomicronemic patients (type V hyperlipoproteinemia) do not have LPL deficiency.

Wondering if the apoprotein activator(s) of LPL might be missing from the chylomicrons of type I patients, we have collected their apoproteins. C-II is

present and, *in vitro*, proves to be capable of stimu-
lating hydrolysis of triglyceride emulsions by LPL.
This disease is therefore not due to absence or alteration in apoprotein C-II.

Further Steps in VLDL Metabolism

At one time it was thought that the only significant difference between VLDL and LDL was the amount of
glyceride present. Very early studies in Berkeley
suggested that VLDL was degraded to smaller and smaller
particles. Gitlin, in the 1950's, was the first to tag
the apoprotein portion in VLDL and re-inject these
lipoproteins intravenously in another subject. The
label soon turned up in LDL, but the reverse, conversion of LDL to VLDL, never occurred. Study of the
metabolic behavior of ^{125}I-tagged apoproteins has become popular again with the exposure of the heterogeneity of the apoproteins. Many such studies have
been done in Bethesda, most lately by Eisenberg, Bilheimer, and Levy. Their data confirm the conversion
of VLDL to LDL. The quantitative aspects of such experiments are tricky and one must be skeptical of
their conclusion that one VLDL particle is converted
to one LDL molecule. All LDL do not necessarily arise
via VLDL, but some undoubtedly do. Only a portion of
the VLDL "micelle" is preserved as it is catabolized,
consistent with it having an LDL core as I speculated
earlier. Even in this "core" some further transforma-
tions may occur during VLDL catabolism. The net result
is that the C-apoproteins (and probably the Fraction
II proteins) are stripped away, along with > 95 per
cent of the triglyceride and > 80 per cent of cholesterol and phospholipids. Apparently the cholesteryl
ester content of the residual LDL is also enriched at
the expense of phospholipid and "free" (unesterified)
sterol.

LCAT

This last change is believed to be the handiwork
of a plasma enzyme, lecithin-cholesterol acyl trans-

ferase (LCAT), sometimes called the Glomset enzyme, since he did the extensive work required to define the "cholesteryl esterase" shown long ago to be in plasma. LCAT catalyzes the transfer of an acyl group from the β position of lecithin to the 3-OH group of cholesterol. Nichols and others have postulated that a major role is assigned this enzyme in the metabolism of VLDL, *i.e.*, as glycerides are peeled off by hydrolysis, the remnant left in the blood is stabilized by replacing the missing triglyceride with another nonpolar ester, cholesteryl ester. The supply of steryl ester is limited and is augmented by esterification of free sterol through the LCAT reaction. The other product, lysolecithin (lecithin shed of one of its two acyl groups), appears to bind to albumin or other proteins and move at least temporarily out of the lipoprotein spectrum. Proofs are still needed as to whether the LCAT-catalyzed reaction, an entirely intravascular event that proceeds at a relatively sluggish rate compared to the para-vascular hydrolysis of triglycerides, occurs intramolecularly in the same lipoprotein or involves lipids in different lipoproteins, and whether its "stabilization role" is the major function of this enzyme. Shortly we will discuss another apoprotein that seems to regulate the activity of LCAT.

Another Mutant Affecting VLDL and LDL.

During the conversion of VLDL to LDL, fleeting intermediates have been observed as the fate of labeled lipoproteins has been traced. These intermediate lipoproteins resemble others that accumulate in certain patients with a peculiar form of hyperlipidemia. Certain of their lipoproteins float in the ultracentrifuge at density 1.006 (and thus are VLDL, according to the rules) but they have the β-motility of LDL on paper or starch electrophoresis. This "floating beta" is characteristic of type III hyperlipoproteinemia. The latter is sometimes familial; at other times it seems to be secondary to hypothyroidism, severe diabetic acidosis, or liver disease. The phenomenon suggests that some normal step in the degradation of VLDL is arrested. In the familial cases there is evidence that LPL activity is normal and that the apoB in the aberrant lipoproteins is at least immunochemically identical to the normal. "Type III" is a provocative problem under heavy study at present. Most likely an enzyme-catalyzed (intracellular) reaction is defective, but no such normal step in VLDL catabolism has yet been elucidated.

Apoprotein Transfers

It has long been known that certain lipids, particularly phospholipids and free cholesterol, presumably being more accessible to the oil-water interface than more hydrophobic lipids, quickly transfer between lipoproteins in different families, between lipoproteins and red cells, and doubtless--although not proved--between lipoproteins and other tissue sites. With availability of labeled apoproteins and better separation techniques for nonlabelled ones, it appears that there is also exchange of certain apoproteins between lipoprotein classes. This is most obvious for the C-proteins, which shuttle between chylomicrons or VLDL and the HDL family. Movement occurs in either direction. All of the C-proteins seem to move more-or-less together; what complement of lipid they may carry with them is not known.

High Density Lipoproteins

HDL dominate the lipoprotein spectrum in most species. In the dog, for example, about 90 per cent of the plasma lipids are in HDL. I have summarized reasons to believe that chylomicrons, VLDL and LDL represent a syndicate mainly engaged in triglyceride transport. HDL also seem to participate in this process, but in a less direct fashion, and there must be other functions for this lipoprotein family. There is a mutant disorder, Tangier disease, in which HDL cannot be sustained, to help us speculate what such functions might be.

The HDL are lipoproteins that float between density 1.063 and 1.210 g/ml. They are spherical macromolecules of a diameter that ranges between 80–140 Å. When negatively stained, they appear under the electron microscope like those shown in Fig. 1. Examination--perhaps it is over-reading--of such micrographs has suggested the existence of substructure, possibly representing aggregation of four or five or more lipid and protein subunits. These concepts must be reconciled with the earlier mentioned view of HDL as a single micelle. The calculated molecular weights of HDL range from about 170,000 to 400,000. From the early days of ultracentrifugal fractionation, it has been conventional to divide HDL into two major subclasses: HDL_2 (mean hydrated density 1.09) and HDL_3 (mean hydrated density 1.14). S_f units were not assigned HDL in the original ultracentrifugal method of Gofman and Lindgren, which was carried out at density 1.063. The flotation rates under standard conditions at density 1.20 ($F_{1.20}$) assigned HDL_2 are between 3.5 and 9, and for HDL_3 lie between 0 and 3.5. A small HDL_1 fraction with a mean hydrated density of 1.05 gm/ml is poorly understood and usually avoided in both collections and discussions of HDL. Beyond the traditional density limits of HDL, small amounts of lipoprotein are insoluble, usually between densities 1.210 and 1.215 g/ml. These have been called very high density lipoproteins (VHDL). They principally contain lysolecithin, at least some of which is probably the product of the LCAT reaction. The amounts of HDL_2 and HDL_3 vary independently, suggesting mixed populations under diverse control; but divisions between density subclasses is arbitrary, and significant chemical differences are perceptible only at the outer extremes of the HDL distribution. The HDL apoproteins have mainly been studied in pools containing both HDL_2 and HDL_3.

The composition of HDL shown in Fig. 10 is a rough average for human HDL_2 + HDL_3. Molecules of higher density have proportionately more protein and less lipid and vice versa. The lipid consists mainly of cholesteryl esters and phospholipid, (there being about 4 mols

247

COMPOSITION OF HIGH DENSITY LIPOPROTEINS

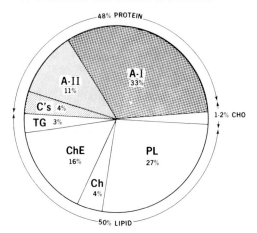

Fig. 10. Representation of the "average" composition
of HDL. For explanation of abbreviations
see legend of Fig. 4 and text.

of phosphatidyl choline per mol of sphingomyelin, a
ratio distinctly different from that in LDL).

HDL Apoproteins.

Until relatively recently it was assumed that
apoHDL consisted of a single protein subunit of molecular weight 21,000 to 31,000, although heterogeneity had
been suggested by immunochemical studies and partial
fractionations. In 1968 the Shores purified and partially characterized two different principal components
in apoHDL. They reported one protein as containing
carboxyl-terminal threonine, the other carboxyl-terminal
glutamine. The "apoLP-thr" in the literature, however,
must now read apo-gln-I, or simply A-I; the original
analyses have proved to be in error. Thus, you will
note in Table 2 that both major HDL apoproteins have
carboxylterminal glutamine.

Many methods are used to isolate apoproteins in
HDL. The most commonly used combine gel filtration
and ion exchange chromatography to obtain maximum purity

248

(Fig. 11). The yield of proteins from delipidated human HDL is as follows: the greatest mass (65 to 75 per

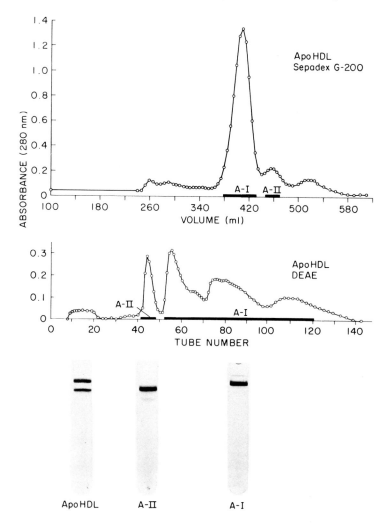

Fig. 11. Separation of delipidated HDL apoproteins by gel filtration (top), or by DEAE ion exchange chromatography (middle). The multiple peaks containing A-I on DEAE all migrate to the same position after polyacrylamide gel electrophoresis (below). The peak following A-II on Sephadex contains the C-proteins.

cent) is apoA-I; A-II represents about 20-25 per cent, and about 5 per cent is accounted for by C-I, C-II and C-III. The relative proportions of the C-proteins appears to be about the same as in apoVLDL, but the limits of variation have not been set. Normally, the mono- and di-sialylated forms of apo-C-III are both present. There remain small amounts of other proteins, including members of the "Fraction-II" group of apo-VLDL. These include the Shores' "R-X$_2$," which is rich in hydrophilic residues, probably "R-X$_1$," and several others.

Apo-A-II.

Discovery of the smaller of what are called the "major" HDL apoproteins was delayed for some time due to the presence of a "blocked" amino-terminus, insusceptible to attack by either the dansylation procedure or the Edman technique. After digestion with pyrrolidoncarboxylyl peptidase, pyrrollidone carboxylic acid (pyroglutamic acid) is recovered.

The structure of A-II is unusual, perhaps unique, and was difficult to establish. As isolated from HDL, the protein appears to have a molecular weight of about 17,000 and contains two half cystines. Lux and colleagues consistently found that, after reduction and treatment with iodoacetic acid, a single protein is obtained which appears to be homogeneous by polyacrylamide gel electrophoresis, analytical gel-isoelectric focussing, or ion exchange chromatography. Multiple forms obtained by Scanu and coworkers appear now to have been most likely due to carbamylation. The reduced and alkylated "monomer" contains one carboxymethyl cystine residue and is immunochemically distinct from any of the other known apoproteins. Although it is difficult to rule out all possiblity that sulfhydryl bonding occurs during purification procedures, it is highly probable that the A-II in circulating HDL consists of two identical monomers (Fig. 7) joined by a single disulfide bond six residues in from the amino-terminal residues.

Comparisons with other Apoproteins. The complete amino acid sequence of the monomer (calculated molecular weight 8690) was determined in Bethesda a few months ago by Brewer, Lux, Ronan, and John (Fig. 7). A-II contains no carbohydrate; it also lacks histidine, arginine and tryptophan (Table 3). Like C-I, it is relatively rich in lysine (9/77 residues). Other frequently occurring amino acids are glutamic acid, glutamine, and threonine. As with the other two apoproteins in Fig. 7, inspection of the primary structure does not reveal any lengthy hydrophobic segments. In five positions along the monomer chain, lysine is adjacent to a dicarboxylic amino acid; but the frequency of such acid-base conjunctions is perhaps not significantly greater than that expected in random assortment within a chain of the same amino acid content. Again, it is probable that the conformation of this apoprotein determines its capacity for lipid-binding. In the case of A-II, this capacity is considerable, for it avidly recombines with phospholipid *in vitro*. The amount of α-helix in delipidated apo-gln-II increases upon reassociation with lipid. This lipid-binding capacity is not decreased by reduction and alkylation although loss of the sulfhydryl bridge does seem to decrease the amount of α-helix in the delipidated apoprotein.

Genetic Relationships. Barker and Dayhoff have compared the primary structures of A-II and C-III and find a high probability that they are distantly related proteins (Table 4). When the entire sequence of the shorter C-protein, C-I, is compared with the first 59 residues of these two other apoproteins, the genetic relationship of C-I to A-II appears to be at least as strong as that previously described between C-I and C-III. On the other hand, the early sequences of A-I (see below) and A-II are not similar. Such data may come to haunt the ABC nomenclature system and indicate why the sorting of apolipoproteins into arbitrary groups is a hazardous base for a permanent nomenclatural system.

ApoA-I.

The first major peak obtained when apoHDL is separated by gel filtration (Fig. 11) contains A-I (the old "R-thr"), an apoprotein for which a complete amino acid sequence has not yet been published. There are also some unresolved questions concerning its homogeneity. The protein can be isolated by gel filtration free of any contamination with A-II and the C-apoproteins (Fig. 11). A-I gives a single broad band on polyacrylamide gel electrophoresis at alkaline and acid pH. "Polymorphic forms" of A-I are obtained by DEAE chromatography or with isoelectric focusing; but these different proteins have the same amino acid composition, are immunochemically identical, have the same migration on polyacrylamide gel, and the same molecular weight (about 27,000). There is no cysteine or cystine in A-I. There appears to be no sialic acid or hexosamines in A-I although undefined neutral sugars may be present. The microheterogeneity of A-I may be due to carbamylation or to changes in amide content occurring during fraction.

Partial Sequence. Partial sequences of proteins analagous to A-I derived from man, bull, and chicken are available in preliminary form (Fig. 12). These include the first 39 amino acids of the human apoprotein reported by the Shores. Analyses of material from the bull and the chicken, reported by R.S. Levy and co-workers, were performed on partially fractionated apoHDL. Judged from the available data, the first portions of A-I from the three species appear to be quite similar. This comparison is a hazardously incomplete venture into the comparative biochemistry of apoHDL, but the apparent homology of the major HDL apoproteins in the rat, human, and dog, as suggested by polyacrylamide gel patterns and other data, indicate that the structure and presumably function(s) of HDL in man were well established millions of years prior to his appearance.

ApoLp-Gln-I

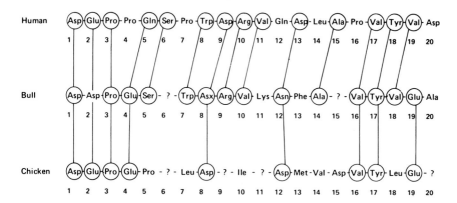

Fig. 12. Comparison of the reported sequences of the first 20 amino acids in human A-I (ApoLP-Gln-I) and homologous proteins in the bull and the chicken. (Data from B. Shore and V. Shore: Proc. European Soc. Atheroscl. Masson and Cie, Paris, France, in press; R.S. Levy and M.V. Martin: Fed. Proc. 30: 788, 1971; R.S. Levy et al.: Fed. Proc. 31: 3776, 1972).

Apoprotein Separatism.

The proportions of A-I and A-II in HDL normally seem to be quite constant. There is one natural experiment in which the proportions have been badly altered, and the presumed result is a near absence of circulating HDL. This is Tangier disease, recognized most easily by the orange tonsils of the patients-- a marker resulting from phenomenal storage of cholesteryl esters in reticuloendothelial tissues. The homozygote for this rare gene has only traces of circulating HDL, called HDL_T because it is immunochemically distinguishable from the normal. As shown in Fig. 13, HDL_T is disproportionately short of apoA-I. The ratio of A-I to A-II is about 1 to 12 (w/w) instead of the normal 3 to 1. The "minor protein" peak in Fig. 13

253

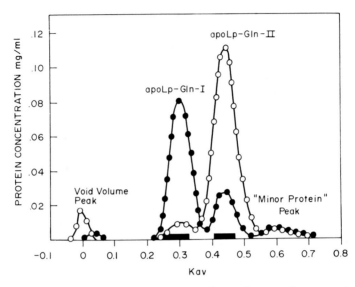

Fig. 13. Sephadex chromatography of equal amounts of
HDL apoproteins from normal subject •——•
and patient with Tangier disease o——o
ApoLP-Gln-I = A-I, apoLP-Gln-II = A-II, the
"Minor Proteins" = C-proteins. Reprinted
from S.E. Lux *et al.*: J. Clin. Invest. 51:
2505, 1972.

contains C-I, C-II, C-III, and C-III-2. Each is present in HDL_T. They represent their usual proportion
(5 to 7 per cent) of the total HDL proteins, but the
total quantity of these C-proteins is also greatly
reduced. The small amounts of A-I, A-II and of the C
proteins in HDL_T appear to be immunochemically identical to their normal counterparts, but this has not yet
been proved by chemical analyses.

We have speculated that the primary genetic defect
in Tangier disease involves the synthesis of apoprotein
A-I. The story may yet prove to be otherwise, and before it is completed, this mutation may end up revealing much more about the lipoprotein puzzle. What, for
example, are the possible interrelationships between
the major HDL(A) proteins and the C-proteins? As we

saw earlier (Table 4), A–II and C–I seem quite probably
to have arisen from a common ancestral protein. Barker
and Dayhoff have compared the preliminary sequences
available for the first 39 residues of A–I and do not
see evidence for such close relationship between A–I
and A–II.

Although A–I and A–II co-exist in the bulk of HDL
in constant proportions, A–I can be easily separated
from these complexes in a high ultracentrifugal field,
and investigators using immunochemical methods have
reported finding "heavy" HDL of low $F_{1.20}$ which contain
only apoprotein A–I.

A very rough estimate suggests that over half of
the total plasma concentration of C-proteins, in the
fasting state, is in the HDL. Turnover studies have
shown C-proteins moving to HDL as their VLDL carriers
are being catabolized. This has encouraged the concept of HDL as a C-protein repository. But other studies, already briefly mentioned, show C-proteins leaving
HDL to join incoming VLDL and chylomicrons. Given the
genetic similarities between some C-proteins and A–II,
it may be that they are secreted together, as HDL, and
the latter lipoproteins are thus the primary source
rather than the convenient "dump" of C-proteins required for normal glyceride turnover.

What we know of triglyceride metabolism in Tangier
disease is both provocative and inadequate. Chylomicrons and VLDL in these HDL-deprived patients have abnormal lipid composition. The VLDL have an abnormal
migration on electrophoresis reminiscent of the "floating beta" in Type III hyperlipoproteinemia. Triglyceride concentrations are high and there are suggestions
of retarded clearance. What is lacking are data on the
apoprotein content and of turnover in lipoproteins
other than HDL in bearers of the Tangier gene.

Perhaps these data will also lead to an explanation of the profound tissue storage of cholesteryl
esters in Tangier disease. We have found adequate
activity of lysosomal (acid) cholesteryl ester hydrolase in the disorder, and local cholesterol synthesis
does not seem to be abnormally high. It would best

suit the data if the input of cholesteryl esters into
scavenger (RE) tissues were greatly increased through
uptake of lipoproteins too unstable for survival in
plasma. There are several possible bases for unsta-
bility of triglyceride-rich particles in Tangier disease. We have already speculated on a possible short-
age of C-proteins attending HDL deficiency. In addition, it has been shown that A-I is an activator of
the LCAT reaction, also postulated to be of importance
to orderly catabolism of chylomicrons and VLDL. Alternatively, there may also be something that HDL or apoprotein A-I, in particular, lends to the effectiveness
of cholesterol transfer between plasma and the cell
membrane. The answer is still buried in the enigma of
HDL.

Summation

In higher forms of life, specialized systems exist
for the inter-organ transportation of lipids, substances
which share a general property of relative insolubility
in water. Other than free fatty acids, lipids are
transported in combination with certain proteins (apolipoproteins) in a manner providing dispersions in water
that are at once stable and yet easily handled at their
destination. The plasma apolipoproteins seem peculiarly
well-adapted for this purpose and, while comparisons
between species have been limited, it is evident that
the homologs of the human apolipoproteins must have
appeared fairly early in evolution. What I have recon-
structed here concerning the properties, functions, and
interactions of apolipoproteins is both tentative and
flawed, for there are many gaps in essential knowledge.
I have concentrated mainly on only one aspect of lipid
transport, the secretion of endogenous triglycerides
and events that follow soon after their release into
the blood.

We have considered as the first model the liver
parenchymal cell. Here, when net synthesis of glyceride begins, or when these lipids have accumulated in
some critical amount, several apoproteins are also produced. One of these is apoLDL or the B protein, whose

size and structure are still uncertain. Lipid-free apoB is insoluble in intracellular fluids and may be a membranous protein. It probably is synthesized in close proximity to lipids, and it may even be a regulator of lipid synthesis. The B protein is considered essential for secretion of triglycerides.

Concomitant with the production of apoLDL, several other apoproteins are newly synthesized in the liver. Some are still called "Fraction-II" proteins and have not been well characterized, except for identification of R-X$_1$, a protein rich in arginine. Three other apoproteins that are important components of triglyceride-rich lipoproteins have been relatively well-defined. They are smaller than apoB or the Fraction II proteins, having molecular weights of from 7,000 to 10,000 daltons. In the nomenclature I have employed for this discussion, they are known as C-I, C-II, and C-III, or collectively as the C-proteins. The primary structures of C-I and C-III are known and similar enough to suggest derivation from a common ancestral gene. Like apoB, the C-proteins probably exist in the cell in the company of lipids. They bind avidly to phospholipids *in vitro* to form "discs" or bilayers as viewed in fixed preparations under the electron microscope. If triglyceride or cholesteryl esters are added to phospholipids during recombination experiments, the complexes have spherical shape and size of VLDL. Whether the C-proteins each independently form lipoproteins of similar density or are integral parts of the same complex is not yet known. It presently seems more likely that they bind in concert to the periphery of triglyceride-rich lipoproteins, and also to HDL, a location allowing them ready exchangeability and performance of important functions at the oil-water interface of the lipoproteins.

The triglycerides, cholesterol, phospholipids, and apoproteins collectively become visible as "liposomes" in the endoplasmic reticulum and Golgi apparatus. It may be within the Golgi bodies that transferases add sialic acid-containing glycosidic residues to certain apoproteins, most notably, apoB and apoC-III. In the

Golgi bodies, lipoprotein assembly appears to be complete, and droplets of lipoproteins, 300-800 Å in diameter, are discernible that chemically resemble plasma very low density lipoproteins (VLDL). Their next destination is the lymph or blood stream.

In the mucosal cells located in the villi of the small intestine, the processing of glycerides differs in several details. In the intestinal Golgi apparatus, two populations of triglyceride-rich particles are visible. One of these consists of droplets that are about the same size and have a composition very much like the VLDL in liver Golgi bodies. Apart from these stand other groups of larger particles (> 1000 Å) awaiting shipment. These are chylomicrons, containers or newly reassembled glycerides and small portions of phospholipids, cholesterol, and other fat soluble substances that have recently arrived from the intestinal lumen. ApoB is present in both VLDL and chylomicrons. As judged from chylomicrons and larger VLDL isolated in plasma, the quantity of protein per mass of glyceride, and the proportion of the total protein represented by apoLDL, is less than it is in smaller VLDL particles. There seems to be another difference between liver and intestine. As the intestinal cell packages triglycerides and proteins, apoB and some of the "Fraction II" apoproteins are synthsized by the cell; but the C-proteins apparently are not, and are acquired by the chylomicrons and intestinal VLDL as these particles exit from the cell. These C-proteins are at least partly obtained by transfer from the circulating high density lipoproteins (HDL).

As chylomicrons and VLDL move out of the Golgi into the blood stream, new events quickly transpire. Some particles immediately acquire a richer complement of C-proteins "loaned" from HDL. There is something purposeful in this welcoming embrace, for at least one of the C-proteins, C-II, is an "activator" of the enzyme lipoprotein lipase. As the particles come in contact with this enzyme in the endothelial wall of capillaries, the triglyceride is progressively stripped away. Shortly, the VLDL and chylomicrons, some of them originally > 1000 Å in diameter and over 100 million daltons in molecular weight, are reduced to "remnants" of > 200 Å diameter and of 1 to 3 x 10^6 molecular weight. The ultimate residuals are low density lipoproteins (LDL) which contain only apoB and a fairly constant proportion of phospholipids and cholesterol, most of the latter being esterified. Gone is nearly all of the original triglyceride, most of it "absorbed" by cells near the site where lipolysis occurred. Gone, too, is most of the original phospholipid and free cholesterol; some of the latter apparently moves after its esterification, to LDL and HDL; but most of the rapidly departing sterol and phospholipids probably disappear in unstable lipoprotein remnants taken up by the liver and other organs. Some of the original C-proteins are not gone, but have transferred, and will apparently spend some time equilibrating between the protective bulk of HDL and new triglyceride-rich lipoproteins entering the plasma to proceed through the same catabolic cycle.

The image of plasma LDL as a slag-pile of debris accumulating from triglyceride transport as conveyed by this reconstruction is probably too narrow if not badly distorted. LDL may possibly arise through several other pathways. There is no evidence, however, that LDL or its constituent apoB re-enters the cycle of triglyceride transport. The sites of LDL disposal and the methods of disassembly are not known. Man, whose LDL concentration exceeds that of all other species, seems to operate close to the limits of the rate of LDL catabolism, perhaps dangerously so, for LDL seems to be an important atherogenic factor. We still need to know more about differences between species in the demand for triglyceride transport, and of alternate routes of LDL production. In their conversion to LDL, VLDL pass through successive stages in which lipoproteins of intermediate density and composition are transiently produced. In one human genetic abnormality, type III hyperlipoproteinemia, such "intermediates" accumulate in plasma, suggesting failure of some unknown mechanism that participates in the dissolution of triglyceride-rich particles.

Among the operations which accompany the hydro-
lytic removal of triglyceride from lipoproteins is a
replacement reaction. Nonpolar lipids like triglycerides help maintain the spherical form and stability of
lipoproteins. Cholesteryl esters can also serve this
function. Cholesterol is esterified in plasma by an
enzyme lecithin: cholesterol acyl transferase (LCAT),
and it has been suggested that this reaction proceeds
in step with lipolysis to assure the continued stability of the lipoproteins produced. The LCAT reaction
is facilitated by HDL, lipoproteins rich in phosphatidyl choline, the β-acyl group of which is transferred to cholesterol. One of the apoproteins of HDL,
Apo-A-I, also activates LCAT *in vitro*.

HDL constitute the bulk of circulating lipoproteins in many species. Their structure and functions
are still unknown. More than a third of the mass of
HDL is contributed by apoprotein A-I. Its constant
companion, apo A-II, is present in about one-third the
mass of A-I. A-II has a molecular weight of approximately 17,000 and has a most unusual structure, consisting of two identical monomers joined by a single
sulfhydryl bridge. From their primary structures, it
is judged that A-II and several of the C-proteins may
have once had a common ancestor and it is possible
that they are still elaborated together, for about half
of the total plasma complement of C-proteins normally
is resident in (or, probably more accurately) on the
surface of HDL. The C's restlessly transfer between
HDL and triglyceride-rich particles, while the A's are
mainly found always in HDL.

We have mentioned Tangier disease as a possible
source of discovery of the relationship between the
several apoproteins and of the role of HDL, for these
lipoproteins are nearly absent from plasma in this
genetic disease. The residual HDL is absolutely de-
ficient in A-I, suggesting that the defect primarily
affects elaboration or utilization of this apoprotein.
But in this rare disease, there also remain to be ex-
plained abnormalities in triglyceride transport and
heavy tissue storage of cholesteryl esters. In Tangier

disease, as in several other human mutants causing freakish changes in plasma lipid concentration, there still lie hidden important secrets of apoprotein purpose and lipoprotein metabolism.

The last sentence perhaps best summarizes the state of knowledge of all apolipoproteins. Yet information about them is being acquired at an ever increasing rate. It seems certain that the unsatisfying two-dimensional models that are all we have today will soon be assuming a third dimension and have far more revealing features. While it is possible that forthcoming revelations will not hold the key to many human diseases, an important area of physiology will be better illuminated.

This lecture summarizes works of many scientists and laboratories, including our own. Specific citations are provided in an expanded form of this lecture and other reviews and papers of special interest cited.

References

Reviews on Lipoproteins:

1. Alaupovic, P., Kostner, G., Lee, D.M., McConathy, W.J. and Magnani, H.N. Peptide composition of human plasma apolipoproteins A, B and C. Exposes Ann. Biochim. Med. <u>31</u>, 145 (1972).

2. Brockerhoff, H. Lipolytic enzymes. In <u>Advances in Chemistry</u>, American Chemical Society, in press.

3. Fredrickson, D.S. Plasma lipoproteins and apolipoproteins. In <u>Harvey Lectures</u>, Academic Press, New York and London, 1974.

4. Fredrickson, D.S., Gotto, A.M., and Levy, R.I. Familial lipoprotein deficiency (abetalipoproteinemia, hypobetalipoproteinemia, and Tangier disease), Chapter 26. In <u>The Metabolic Basis of Inherited Disease</u>, J.B. Stanbury, J.B. Wyngaarden, and D.S. Fredrickson (Eds.), McGraw-Hill, New York, p. 493, 1972.

5. Fredrickson, D.S., Lux, S.E. and Herbert, P.N. The apolipoproteins. Adv. Exp. Med. Biol. <u>26</u>, 25 (1972).

6. Havel, R.J. Mechanisms of hyperlipoproteinemia. Adv. Exp. Med. Biol. <u>26</u>, 57 (1972).

7. Nelson, G.J. <u>Blood</u> <u>Lipids</u> <u>and</u> <u>Lipoproteins</u>: <u>Quantitation</u>, <u>Composition</u>, <u>and</u> <u>Metabolism</u>, Wiley-Interscience, New York, 1972.

8. Norum, K.R., Glomset, J.A. and Gjone, E. Familial lecithin: cholesterol acyl transferase deficiency. In <u>The</u> <u>Metabolic</u> <u>Basis</u> <u>of</u> <u>Inherited</u> <u>Disease</u>, J.B. Stanbury, J.B. Wyngaarden, and D.S. Fredrickson (Eds.), McGraw-Hill, New York, p. 531, 1972.

9. Scanu, A.M. Human plasma high density lipoproteins. In <u>Plasma</u> <u>Lipoproteins</u>, R.M.S. Smellie (Ed.), Academic Press, London and New York, p. 29, 1971.

10. Schumaker, V.N. and Adams, G.H. Circulating lipoproteins. Ann. Rev. Biochem. <u>38</u>, 113 (1969).

11. Scow, R.O., Hamosh, M., Blanchette-Mackie, E.J. and Evans, A.J. Uptake of blood triglyceride by various tissues. Lipids <u>7</u>, 497 (1972).

12. Shore, B. and Shore, V. Structure of normal and pathological lipoproteins. Exposés Ann. Biochem. Méd. <u>31</u>, 3 (1972).

Special Articles

1. Barker, W.C. and Dayhoff, M.O. Computer studies of distantly related sequences. Biophysical Society Abstracts, Seventeenth Annual Meeting, February 27-March 2, 1973, p. 205a.

2. Barker, W.C. and Dayhoff, M.O. Detecting distant relationships: computer methods and results. In <u>Atlas</u> <u>of</u> <u>Protein</u> <u>Sequence</u> <u>and</u> <u>Structure</u>, Vol. 5,

M.O. Dayhoff (Ed.), The National Biomedical Research Foundation, Silver Spring, Maryland, p. 101, 1972.

3. Eisenberg, S., Bilheimer, D.W., Lindgren, F.J. and Levy, R.I. On the metabolic conversion of human plasma very low density lipoproteins. Biochim. Biophys. Acta, in press.

4. Kostner, G. and Holasek, A. Characterization and quantitation of the apolipoproteins from human chyle chylomicrons. Biochemistry 11, 1217 (1972).

5. Mateu, L., Tardieu, A., Luzzati, V., Aggerbeck, L. and Scanu, A.M. On the structure of human serum low density lipoprotein. J. Mol. Biol. 70, 105 (1972).

6. Shipley, G.G., Atkinson, D. and Scanu, A.M. Small-angle X-ray scattering of human high-density lipoproteins. J. Supramolecular Structure 1, 98 (1972).

7. Singer, S.J. in Structure and Function of Biological Membranes. L.I. Rothfield (Ed.), Academic Press, New York, p. 145, 1971.

8. Windmueller, H.G., Herbert, P.N. and Levy, R.I. Biosynthesis of lymph and plasma lipoprotein by isolated perfused rat liver and intestine. J. Lipid Res. 14, 215 (1973).

263

SUBJECT INDEX

A

ABC code of apolipoprotein
nomenclature, 229–230
Abetalipoproteinemia, 240
ADP, effect on glutamic dehydrogenase,
47
Affinity chromatography of antibodies,
88–90
Alkaline phosphatase induced by steroid
hormones, 208–209
Alpha-chymotrypsin, active site of, 29–32
Alpha-lactalbumin helix profile, 19–20, 22
Alpha-naphthol, 117
Antibodies, 73-107
definition of, 71
evidence for conformational specificity
of, 95–97
heterogeneity in binding affinities of, 76
homogeneity in binding affinities of, 76
immunospecificity of, 88–90
inactivation of staphyloccal nuclease by,
90–93, 96–97
kinetic and equilibrium constants for
reaction with staphylococcal
nuclease, 93–95
lattice theory of precipitation, 76–78
made against apomyoglobin, 82–86
made against myoglobin, 82–86
made against polypeptides, 80, 82–103
made against staphylococcal nuclease,
86–103
Ouchterlony double diffusion analysis
of, 80
precipitin reaction of, 78–80
purification by immunoadsorbent
chromatography, 88–90
Antigens, 73-107

B

Beta-turns, 23–24

ApoA-I, 252
ApoA-II, 250–251
ApoB, 236–237, 239–240
ApoC-I, 232
ApoC-II, 236
ApoC-III, 232
ApoLDL, 236–237
Apolipoproteins, 219–263
A-I, 252
A-II, 250–251
acid-base pairs in, 234–235
B, 236–237, 239–240
C-I, 232
C-II, 236
C-III, 232
comparative studies of, 238–239,
252–253
genetic relationships between, 235–236,
251–252
HDL, 248–256
in metabolism of VLDL, 239–244
LDL, 236–237
nomenclature of, 228–230
primary and secondary structure of,
233–235, 252–253
separatism in, 253–256
transfer of lipids between, 246
VLDL, 230–232, 237–239
Apolipoprotein synthesis, 239
ApoVLDL, 230–232, 237–239
Arene oxides, 114–116, 122–125
Aromatic hydroxylation, mechanism of,
109–133

Eukaryotic gene expression, 187–217
Eukaryotic gene regulation, models for,
177–179
Extended structures in proteins, 25–26
Eukaryotes, 136

F

Fenton's model system for oxidants, 124
Free fatty acids (FFA), 225
Flow dichroism of chromatin, 169–170

G

Gamma-conformation, 5
Gel filtration fractionation of chromatin,
146–148
Gene activity regulation in cells, 135–136
Gene expression in animal cells, 187–217
Gene expression regulation,
at cellular level, 190
at cellular organelle level, 190
at chromosomal level, 189–190
at level of tRNA, 197–199
at molecular level, 188–189
at multicellular level, 190
at transcriptional level, 196–197
cyclic AMP in, 199
models for, 177–179
protein metabolism in, 200
steroid hormones in, 200–211
Genetic relationships between
apolipoproteins, 235–236
Glomset enzyme, 244–245
Glycols, 114–115
Glutamic dehydrogenase, 43–72
cross-linking of, 62–68
effect of toluene on, 58–62
electron microscopy of, 51–54
equilibrium sedimentation studies of,
54–56
light scattering measurements of, 48–51
linear association model of, 58–59
molecular weight of, 48–51
number average molecular weight of,
55–56
radius of gyration determination of,
58–59
reaction catalyzed by, 45
reduced specific viscosity of, 60–61

reversible polymerization of, 45–47
self-assembly of, 44–48
small angle X-ray scattering results for,
59
structure of, 43–72
subunit structure of, 50–51
weight-average molecular weight, 54–56
Z-average molecular weight of, 55–56
Glutathione, 114–116
Glutathione S-epoxide transferase,
114–116
GTP, effect on glutamic dehydrogenase,
47, 60, 66

H

HDL, 220–224, 246–256
HDL apolipoproteins, 248–256
Helix-breaking residues, 10–15
Helix-making residues, 10–15
Helix probability profiles, 15–21
Heparin-displaced lipase activity, 243
Heterochromatin, 154
Heterogeneous nuclear RNA, 194–195
High density lipoproteins, 220–224,
246–256
Histone proteins of chromatin, 156–165,
191–192
Histones, 137
Hydrogen bonds in amino acids, 12–13
Hydrolases and lipoproteins, 224
Hyperlipoproteinemia, 243–244, 245–246
Hypertriglyceridemia, 243–244

I

Immunoadsorbent method of preparing
antibodies, 88–90
Immunogenicity, 73
Ion exchange chromatography in
fractionation of chromatin, 148–149
Ising model, 15–17

K

$^\kappa$conf.
definition of, 96–98
measurement of, 99–101

X

X-ray diffraction patterns of chromatin, 169

X-ray scattering, results for glutamic dehydrogenase, 59

X-ray structure refinement, 21–22

Z

Z-average molecular weight of glutamic dehydrogenase, 55–56

Zimm-Bragg parameters, 13–17

NOTES

NOTES

NOTES

NOTES

NOTES